Mike Hulme
Norwich
March 1990

Ozone Depletion

Reproduced with permission of Dr and Mrs Robin Russell Jones

Ozone Depletion: Health and Environmental Consequences

Edited by

ROBIN RUSSELL JONES
*St John's Hospital for Diseases of the Skin
London*

and

TOM WIGLEY
*Climatic Research Unit, University of
East Anglia, Norwich*

JOHN WILEY & SONS
Chichester · New York · Brisbane · Toronto · Singapore

Copyright © 1989 by John Wiley & Sons Ltd.
Baffins Lane, Chichester
West Sussex PO19 1UD, England

All rights reserved.

Distributed in the United States of America, Canada and
Japan by Alan R. Liss Inc., 41 East 11th Street, New York,
NY 10003, USA.

No part of this book may be reproduced by any means,
or transmitted, or translated into a machine language
without the written permission of the publisher.

Other Wiley Editorial Offices

John Wiley & Sons, Inc., 605 Third Avenue,
New York, NY 10158-0012, USA

Jacaranda Wiley Ltd, G.P.O. Box 859, Brisbane,
Queensland 4001, Australia

John Wiley & Sons (Canada) Ltd, 22 Worcester Road,
Rexdale, Ontario M9W 1L1, Canada

John Wiley & Sons (SEA) Pte Ltd, 37 Jalan Pemimpin 05-04,
Block B, Union Industrial Building, Singapore 2057

Library of Congress Cataloging-in-Publication Data:

Ozone depletion: health and environmental consequences/edited by
  R. Russell Jones and T. Wigley.
      p.   cm.
  'Proceedings of an International Conference on the Health and
Environmental Consequences of Stratospheric Ozone Depletion, held at the
Royal Institute of British Architects, London, on November 28–29,
1988'—Pref.
  Includes bibliographical references.
  ISBN 0 471 92316 8
  1. Ozone layer depletion—Health aspects—Congresses. 2. Ozone layer
depletion—Environmental aspects—Congresses. 3. Greenhouse effect,
Atmospheric—Congresses. 4. Chlorofluorocarbons—Congresses. 5. Global
warming—Congresses. I. Russell Jones, Robin. II. Wigley, T. (Tom)
III. International Conference on the Health and Environmental
Consequences of Stratospheric Ozone Depletion (1988 : Royal Institute of
British Architects)
RA569.8.O96 1989
616.9′88—dc20   89-16614
                                                                  CIP

British Library Cataloguing in Publication Data:

Ozone depletion: health and environmental consequences.
  1. Stratosphere. Ozone. Environmental aspects
  I. Jones, R. Russell   II. Wigley, T. M. L.
  551.5′142

ISBN 0 471 92316 8

Typeset by Proteus, Worle, Avon
Printed and bound by Biddles Ltd., Guildford, Surrey.

Contents

Preface vii

A Fact Sheet about Ozone xi

Contributors to this Volume xv

Contributors to the Conference xvii

PART 1 INTRODUCTION

Chapter 1   Opening Remarks 3
            Lord Zuckerman
Chapter 2   Protecting the Ozone Layer: A Challenge for the
            World Community 5
            Virginia Bottomley

PART 2 OZONE DEPLETION

Chapter 3   The Beginnings of a Problem 15
            Ivar S. A. Isaksen
Chapter 4   Numerical Modelling of Ozone Perturbations 27
            R. S. Eckman and John A. Pyle
Chapter 5   Present State of Knowledge of the Ozone Layer 43
            Robert T. Watson
            Discussion Period 1 59

PART 3 GLOBAL WARMING

Chapter 6   Opening Remarks 69
            John Maddox
Chapter 7   The Greenhouse Effect 71
            Keith P. Shine
Chapter 8   Measurement and Prediction of Global Warming 85
            Tom M. L. Wigley
Chapter 9   Vulnerable Ecosystems 99
            Michael Oppenheimer
            Discussion Period 2 105

PART 4 INTERNATIONAL CONTROLS

Chapter 10 The European Dimension 109
Stanley Clinton Davis

Chapter 11 The Montreal Protocol on Substances that Deplete the Ozone Layer: Its Development and Likely Impact 115
Peter Usher

Discussion Period 3 141

PART 5 ULTRAVIOLET-INDUCED CARCINOGENESIS

Chapter 12 Photosensitive Human Syndromes and Cellular Defects in DNA Repair 147
C. F. Arlett and J. Cole

Chapter 13 Experimental Photocarcinogenesis 161
Jan C. van der Leun

Chapter 14 Epidemiology of Melanoma: Its Relationship to Ultraviolet Radiation and Ozone Depletion 169
J. Mark Elwood

Discussion Period 4 191

PART 6 GLOBAL CONSEQUENCES

Chapter 15 Effects of Ultraviolet-B Radiation on Terrestrial Plants and Marine Organisms.............................. 197
Robert C. Worrest and Lester D. Grant

Chapter 16 Consequences for Human Health of Stratospheric Ozone Depletion 207
Robin Russell Jones

Discussion Period 5 229

PART 7 POLITICAL ASPECTS

Chapter 17 Alternatives to CFCs 235
C. E. Tane

Chapter 18 Environmental Imperatives 243
Jonathon Porritt

Chapter 19 Ozone Depletion: Consumer Choice 251
Rachel Waterhouse

Chapter 20 Priorities for Research 257
D. A. Warrilow

PART 8 CONCLUSION

Chapter 21 Concluding Remarks 267
Lord Zuckerman

Index 273

Preface

This volume represents the edited proceedings of an International Conference on the Health and Environmental Consequences of Stratospheric Ozone Depletion held at the Royal Institute of British Architects, London, on 28–29 November 1988. The Conference was chaired by Lord Zuckerman, OM, KCB, FRS, formerly chief scientific adviser to the Cabinet, and organized by Dr Robin Russell Jones, MRCP, Consultant Dermatologist at St John's Hospital for Diseases of the Skin, London. The Conference was sponsored by a number of organizations, principally the European Commission, via the Europe Against Cancer Campaign, the Consumers Association, and Friends of the Earth UK. Financial support was also received from the British Antarctic Survey and the Cancer Research Campaign. We would like to express our appreciation to all of these organizations.

In many ways the Conference was unique in bringing together the main participants in the medical, scientific and political arenas. Since the announcement in 1985 by the British Antarctic Survey of a 'hole' in the ozone layer over Antarctica, there had been increasing disquiet amongst the scientific community about the continued use of chlorofluorocarbons (CFCs) and other chemicals that destroy stratospheric ozone, and increasing public awareness of the dangers of ultraviolet light. This concern however, was not met immediately by an appropriate political response. It was not until September 1987 that the United Nations Environment Programme persuaded the majority of the industrialized nations to sign the Montreal Protocol. This agreement called for a cut in consumption of CFCs by the year 2000 of 50%, and a cut in production of only 35%. Shortly afterwards, the results of an international scientific survey in Antarctica became available showing an even greater ozone loss during springtime and providing clear and unmistakable evidence of the primary role played by CFCs in this depletion.

In March 1988 the Ozone Trends Review Panel published their report derived from ground-based observations, showing a modest but significant decline in ozone levels over the northern hemisphere from 1969 onwards. This trend was again consistent with the effect of man-made chemical releases.

It was against this background that the Conference was organized. It was our intention throughout to introduce both scientific and political considerations into the programme, so that politicians could assess the strength of the scientific data, and so that scientists could judge the adequacy of the political response. From this point of view the Conference was an outstanding success. Virginia Bottomley on behalf of the UK Government, Stanley Clinton Davis on behalf of the Commission of the European Communities, and Peter Usher on behalf of UNEP all recognized the urgency of the problem and underlined the need for greater cuts in CFC production and indeed, for their eventual elimination. The Conference attracted considerable media interest, not least because, in the week before the Conference, the British Prime Minister, Margaret Thatcher, announced that the UK Government, under the auspices of UNEP, would host a major conference on ozone depletion in London in March 1989. In addition, ICI announced the construction of two new plants, one in the UK and one in the USA for the production of CFC substitutes.

From a scientific point of view the Conference brought together the acknowledged world experts in different aspects of ozone depletion. Because CFCs are also potent greenhouse gases, a section on global warming was included in the programme to underline the twin dangers posed by this species of chemical. These two issues, stratospheric ozone depletion and global warming, have now assumed such importance internationally that it is doubtful whether such a conference could again be staged by a non-governmental organization and hope to attract such a concentration of expertise. Because the discussion periods served both to clarify the research findings and to highlight areas of outstanding concern, they have been retained in this volume in edited form.

All but one of the papers presented at the Conference has been included in this volume. The exception is the paper by Joe Farman from the British Antarctic Survey, who felt unable to furnish a written version of his oral presentation in time to meet the publisher's deadline.

Delegates at the Conference received an information sheet itemizing the principal scientific and political issues relating to ozone depletion. Because this was helpful for people without specialist knowledge of the subject it has been retained in this volume.

Numerous people have provided generous help in the preparation of this book. Although we cannot thank them all we would particularly like to express our appreciation to the chapter authors who willingly and cheerfully met tight deadlines in order to bring this volume to completion as soon as possible. The speed of preparation of this volume and the heavy commitments of many of the authors have meant that there are stylistic differences between the various chapters, which we trust will not prove irksome to the reader. We owe a considerable debt of gratitude to the many secretaries who coped with a considerable volume of work, and in particular to Barbara Gill who fulfilled the task of conference secretary with patience and efficiency. Finally we would like to thank the other members of the organizing committee, Simon Kleine and Jo Griffiths from the Consumers' Association, and Fiona Weir from Friends of the Earth.

ROBIN RUSSELL JONES and TOM WIGLEY

April 1989

A Fact Sheet about Ozone

This brief synopsis has been prepared for those delegates without specialist knowledge of stratospheric ozone, the chemicals that destroy ozone, and the international agreement to protect the ozone layer.

1. The atmosphere is divided into layers that are defined approximately by distance above the surface of the Earth:

0–15 km	Troposphere
15–50 km	Stratosphere
50–85 km	Mesosphere
> 85 km	Thermosphere

2. A proportion of the Sun's energy is emitted as ultraviolet (UV) radiation or light. Ultraviolet light is divided into three types according to wavelength:

UVA	320–400 nm
UVB	290–320 nm
UVC	< 290 nm

   DNA, the genetic code present in all living cells, is damaged by ultraviolet radiation. UVC is the most damaging, then UVB, then UVA. Damage to DNA can either kill the cell or cause it to mutate and become cancerous.

3. Ozone (O_3) is a gas comprising three atoms of oxygen. It is present in significant amounts in both the stratosphere and the troposphere, the combined amount being known as total column ozone. Stratospheric ozone is generated by the action of ultraviolet light on molecules of oxygen. Thus:

$$O_2 \rightarrow O + O$$
$$O + O_2 \rightarrow O_3$$

   Tropospheric ozone is generated by the action of ultraviolet light on molecules of nitrogen dioxide. Thus:

$$NO_2 \rightarrow NO + O$$
$$O + O_2 \rightarrow O_3$$

4. Formation of tropospheric ozone is accelerated by sunlight acting on atmospheric pollutants, particularly hydrocarbon emissions from car exhausts and NO_x emissions from transport and industry. At high concentration, tropospheric ozone has deleterious effects on human health (WHO recommend an 8 h maximum in air of 60 parts per billion). Tropospheric ozone also reduces crop yields in sensitive species and damages forests. Finally, tropospheric ozone is a potent greenhouse gas and may contribute significantly to the problem of global warming.
5. Stratospheric ozone, on the other hand, plays a critical role in protecting living organisms from the harmful effects of ultraviolet radiation. Ozone is broken down by sunlight, and this process absorbs shorter wavelength ultraviolet light (ultraviolet radiation below 315 nm). Stratospheric ozone therefore acts as a shield around the Earth, which virtually eliminates UVC and greatly reduces the amount of UVB reaching the surface of the planet. Without the protective effect of ozone, terrestrial ecosystems could not have evolved in the way that they have.
6. Concern has arisen because certain man-made chemicals can catalyse the destruction of stratospheric ozone. Principal among these chemicals are chlorofluorocarbons (CFCs) and halons. Both CFCs and halons are stable chemicals which persist for many years in the atmosphere. They eventually break down in the stratosphere to release halogens; chlorine in the case of CFCs, and bromine in the case of halons. Chlorine and bromine can destroy ozone, but are not themselves destroyed by this process. Thus, one halogen atom can destroy thousands of molecules of ozone. The ability of a chemical to destroy ozone depends upon the halogen type (chlorine or bromine), the number of halogen atoms it releases, and its residence time in the atmosphere. Each chemical can be assigned a number according to its ozone depletion potential (ODP).
7. The discovery of a large hole in the ozone layer over Antarctica has prompted political action. The Vienna Convention for the protection of the ozone layer was established in 1985 under the auspices of the United Nations Environment Programme (UNEP). In September 1987, the world's prime consumers and manufacturers of CFCs and halons reached agreement in Montreal to reduce emissions into the atmosphere. The Montreal Protocol covers the main halogenated CFCs and halons of commercial significance. These are listed in Table 1 together with their chemical names, their ozone depletion potential, and their 1985 global production rates.
8. The Montreal Protocol commits signatory nations to a freeze on the consumption and production of group 2 substances at their 1986 level. For group 1 substances the Protocol requires a freeze on consumption at 1986 levels by 1990, a cut to 80% by 1994, and a further cut to 50% by 1999. Production, on the other hand, is allowed to rise by 10% by 1990 before

reducing to 90% by 1994 and 65% by 1999. The reason for the difference between consumption and production is to make allowance for developing and low-consumer countries who, it is argued, might need to import CFCs as their economy develops.

Table 1 Chemicals included in the Montreal Protocol

Substance		Ozone depletion potential*	1985 world production (million kg/year)
Group 1			
CFC 11	Trichlorofluoromethane	1.0	370
CFC 12	Dichlorodifluoromethane	1.0	470
CFC 113	Trichlorotrifluoromethane	0.8	160
CFC 114	Dichlorotetrafluoroethane	1.0	Very low
CFC 115	Chloropentafluoroethane	0.6	Very low
Group 2			
Halon 1211	Bromochlorodifluoromethane	2.7	10
Halon 1301	Bromotrifluoromethane	11.4	10
Halon 2402	Dibromotetrafluoroethane	5.6	Very low

*ODP values from SORG 1988.

9. The Montreal Protocol has not allayed fears about stratospheric ozone depletion. There are several reasons for this.

First the prolonged residence times of chlorine- and bromine-bearing substances means that very much larger cuts in production are needed to stabilize chlorine and bromine levels in the atmosphere. Most experts agree that an 85% cut in production is necessary just to stabilize atmospheric concentrations at their present level.

Second, the appearance each year of the Antarctic hole and the thinning observed over the Northern Hemisphere indicates that the models previously used to predict ozone depletion may underestimate the seriousness of the problem.

Third, a number of ozone-depleting chemicals are not covered by the Montreal Protocol. Principal among these are methylchloroform (ODP 0.15) and HCFC 22 (ODP 0.05). Already methylchloroform is produced in larger quantities than any other ozone-depleting chemical, and the US Environmental Protection Agency has calculated that chlorine levels might treble by the year 2075 if methylchloroform emissions continue to grow at 3.5–5% per annum.

Finally, some countries with independent CFC manufacturing capacity have not signed the Montreal Protocol, or even the Vienna Convention for the Protection of the Ozone Layer. The two largest such countries are China and India. At the present time, UNEP is working towards a revision of the Montreal Protocol. It remains to be seen, however, what new measures will be proposed, which chemicals will be included, and which countries will agree to sign. The European Community, the USA and many other countries have called for a total phase-out of CFCs.

10. In 1986, CFCs 11 and 12 accounted for two thirds of the total ozone-depleting potential of all man-made releases: 30% of production was used in aerosols, 34% in foamed plastics and 30% in refrigeration and air-conditioning systems. CFCs are also used as solvents and degreasing agents in the metal and electronics industries, while halons are widely used as fire retardants.

However, concern about these chemicals is not confined to their effects on stratospheric ozone. CFCs are also potent greenhouse gases — indeed a molecule of CFC 11, 12 or 113 has more than 10 000 times the impact on global warming of a CO_2 molecule. HCFC 22 is also a potent greenhouse gas.

The challenge for industry, therefore, is to find substitute chemicals or substitute technologies which are environmentally benign, do not threaten the ozone layer and do not contribute to greenhouse warming.

ROBIN RUSSELL JONES and TOM WIGLEY

Contributors to this Volume

COLIN F. ARLETT, *MRC Cell Mutation Unit, University of Sussex, Falmer, Brighton, Sussex BN1 9RR, UK.*

VIRGINIA BOTTOMLEY, *Parliamentary Under-Secretary of State, Department of the Environment, Marsham Street, London SW1, UK.*

STANLEY CLINTON DAVIS, *Commissioner of the European Communities, Rue de la Loi 200, 1049 Brussels, Belgium.*

J. COLE, *MRC Cell Mutation Unit, University of Sussex, Falmer, Brighton, Sussex BN1 9RR, UK.*

R. S. ECKMAN, *NASA Langley Research Center, Hampton, Virginia 23005, USA.*

J. MARK ELWOOD, *Professor of Community Health, University of Nottingham, Queen's Medical Centre, Nottingham NG7 2UH, UK.*

LESTER D. GRANT, *Director, Office of Health and Environmental Assessment, US Environmental Protection Agency, Research Triangle Park, North Carolina, USA.*

IVAR S. A. ISAKSEN, *Professor, Department of Geophysics, University of Oslo, PO Box 1022, Blindern 0315, Oslo, Norway.*

JAN C. VAN DER LEUN, *Professor of Dermatology, University Hospital Ultrecht, Heidelberg laan 100, NL-3584 CX, Utrecht, The Netherlands.*

JOHN MADDOX, *Editor of* Nature, *4 Little Essex Street, London WC2R 3LF, UK.*

MICHAEL OPPENHEIMER, *Environmental Defense Fund, 257 Park Avenue South, New York City, NY 10010, USA.*

JONATHON PORRITT, *Director, Friends of the Earth, 26–28 Underwood St, London N1 7JQ, UK.*

JOHN A. PYLE, *Department of Chemistry, University of Cambridge, Cambridge CB2 1EW, UK.*

ROBIN RUSSELL JONES, *Consultant Dermatologist, St John's Hospital for Diseases of the Skin, 5 Lisle Street, London WC2 7BJ, UK.*

KEITH P. SHINE, *Department of Meteorology, University of Reading, 2 Earley Gate, Reading RG6 3AU, UK.*

C. E. TANE, *Marketing Manager, New Fluorocarbons, ICI Chemicals and Polymers Ltd., PO Box 13, The Heath, Runcorn, Cheshire WA7 4QF, UK.*

PETER USHER, *Programme Officer, UNEP, PO Box 30552, Nairobi, Kenya.*

DAVID WARRILOW, *Project Manager, Air Quality Division, Department of the Environment, Marsham St, London SW1, UK.*

RACHEL WATERHOUSE CBE, *Chairman, Consumers' Association, 2 Marylebone Road, London NW1, UK.*

ROBERT T. WATSON, *Chairman, Ozone Trends Review Panel, Chief, Upper Atmosphere Research Program, Division of Earth Sciences and Applications, NASA HQ, 600 Independence Avenue NW, Washington DC, USA.*

TOM M. L. WIGLEY, *Professor, Climatic Research Unit, University of East Anglia, Norwich NR4 7TJ, UK.*

ROBERT C. WORREST, *Manager Stratospheric Ozone Research Program, US Environmental Protection Agency, Washington DC 20460, USA.*

LORD ZUCKERMAN OM, KCB, FRS, *The Cabinet Office, 70 Whitehall, London SW1, UK.*

Contributors to the Conference

The following contributed to the International Conference on the Health and Environmental Consequences of Stratospheric Ozone Depletion held at the Royal Institute of British Architects, London, on 28–29 November 1988.

LIST OF CHAIR PERSONS

Conference Chairman

LORD S. ZUCKERMAN OM, KCB, FRS,
The Cabinet Office.

Sessional Chair Persons

JAMES LOVELOCK, FRS,
Visiting Professor of Cybernetics, University of Reading.

RONA MACKIE,
Professor of Dermatology, University of Glasgow.

JOHN MADDOX,
Editor of Nature.

ROBIN RUSSELL JONES,
Consultant Dermatologist, St John's Hospital for Diseases of the Skin.

BRIAN THRUSH,
Professor of Physical Chemistry, University of Cambridge.

xviii   CONTRIBUTORS TO THE CONFERENCE

LIST OF SPEAKERS

COLIN ARLETT,
MRC Cell Mutation Unit, University of Sussex, Falmer, Brighton, Sussex BN1 9RR, UK.

VIRGINIA BOTTOMLEY,
Parliamentary Under-Secretary of State, Department of the Environment, Marsham Street, London SW1, UK.

STANLEY CLINTON DAVIS,
Commissioner of the European Communities, Rue de la Jui 200, 1049 Brussels, Belgium.

J. MARK ELWOOD,
Professor of Community Health, University of Nottingham, Queen's Medical Centre, Nottingham NG7 2UH, UK.

JOE FARMAN,
British Antarctic Survey, High Cross, Madingley Road, Cambridge CB3 0ET, UK.

IVAR S. A. ISAKSEN,
Professor, Department of Geophysics, University of Oslo, PO Box 1022, Blindern 0315, Oslo, Norway.

JAN C. VAN DER LEUN,
Professor of Dermatology, University Hospital Ultrecht, Heidelberg laan 100, NL-3584 CX, Utrecht, The Netherlands.

MICHAEL OPPENHEIMER,
Environmental Defense Fund, 257 Park Avenue South, New York City, NY 10010, USA.

JONATHON PORRITT,
Director, Friends of the Earth, 26–28 Underwood St, London N1 7JQ, UK.

JOHN A. PYLE,
Department of Chemistry, University of Cambridge, Cambridge CB2 1EW, UK.

ROBIN RUSSELL JONES,
Consultant Dermatologist, St John's Hospital for Diseases of the Skin, 5 Lisle Street, London WC2 7BJ, UK.

KEITH SHINE,
Department of Meteorology, University of Reading, 2 Earley Gate, Reading RG6 3AU, UK.

C. E. TANE,
Marketing Manager, New Fluorocarbons, ICI Chemicals and Polymers Ltd,
PO Box 13, The Heath, Runcorn, Cheshire WA7 4QF, UK.

PETER USHER,
Programme Officer, UNEP, PO Box 30552, Nairobi, Kenya.

DAVID WARRILOW,
Project Manager, Air Quality Division, Department of the Environment,
Marsham St, London SW1, UK.

RACHEL WATERHOUSE CBE,
Chairman, Consumers' Association, 2 Marylebone Road, London NW1, UK.

ROBERT T. WATSON,
Chairman, Ozone Trends Review Panel; Chief, Upper Atmosphere Research
Program, NASA HQ, 600 Independence Avenue NW, Washington DC, USA.

TOM M. L. WIGLEY,
Professor, Climatic Research Unit, University of East Anglia, Norwich NR4 7TJ,
UK.

ROBERT WORREST,
Manager, Stratospheric Ozone Research Program, US Environmental Protection
Agency, Washington DC 20460, USA.

LORD ZUCKERMAN OM, KCB, FRS,
The Cabinet Office, 70 Whitehall, London SW1, UK.

Part 1
Introduction

1
Opening Remarks

LORD ZUCKERMAN

In opening this Conference may I, on behalf of our sponsors, offer a word of welcome to all who are here. Many of you have come from far away, and the fact that a member of Her Majesty's Government is to give the Opening Address is a measure of the importance that is now attached to the subject we shall be discussing during the next two days.

Twenty years ago — perhaps even ten years ago — I do not think that a discussion on the depletion of the ozone layer would have attracted an audience that would have filled even the front two rows of this hall. Today every seat is occupied. Nor was there then much interest in topics such as the greenhouse effect or acid rain. Indeed, the existence of an acid rain problem was even questioned in those days in government circles. Today the importance of all these issues is widely appreciated and understood.

In the knowledge that this is so, may I introduce our first speaker. Mrs Bottomley is the Member of Government in the Department of the Environment responsible for dealing with environmental problems in general, and with the depletion of the ozone layer in particular. Speaking as she does on behalf of the Government, I am quite sure she will make clear the seriousness with which the problem is now treated in official circles.

2
Protecting the Ozone Layer: A Challenge for the World Community

VIRGINIA BOTTOMLEY
Department of the Environment, UK

I appreciate the opportunity of being able to launch the sixth in a series of keynote statements on the environment by DOE Ministers. It is a timely occasion to bring you up to date with the steps the Government is taking to tackle the major global issue of ozone depletion; and to try to bring together the strands of recent developments, both scientific and political — a process in which Britain is playing a leading role.

THE MONTREAL PROTOCOL: A LANDMARK IN ENVIRONMENTAL PROTECTION

It is easy with hindsight to say that the Montreal Protocol, which was adopted last September, should have been tougher. It could have concentrated on controlling production of CFCs and halons rather than consumption. It could have found a better definition of consumption than production plus imports minus exports. It could have been tougher on developing countries. Above all, it could have demanded more in the way of CFC reductions than 50% by 1999.

But in September 1987 it was not universally accepted, apart from the phenomenon of the so-called Antarctic hole, that scientists had produced unambiguous or undisputed evidence of ozone depletion. I think even Joe

---

Ozone Depletion: Health and Environmental Consequences
Edited by R. Russell Jones and T. Wigley
© 1989 John Wiley & Sons Ltd

Farman, whose British Antarctic Survey team discovered the hole and who will be speaking later this morning, will accept that other possible causes than CFCs were still seriously being considered last year. There are several people here today who took part in the Montreal process. They can tell you that the Protocol was adopted only after difficult and prolonged negotiations. The 50% CFC reductions called for represented the utmost that could have been agreed at the time, and there were problems with Third World countries who suspected that they were unfairly treated.

The most important and far-sighted provision in the Protocol is its guarantee of 4-yearly reviews of the science, environmental and health effects, economics and technology, with the first review to be completed in 1990.

The British Government saw the Protocol as a major environmental landmark. It was precautionary. It went further than the undisputed scientific evidence of the day. It was also drafted so as not to frighten off countries that had not taken a full part in the negotiations. We are heartened by the fact that while the UK and 24 other countries signed up in Montreal, the list of signatories is now well over 40. That is still not nearly enough. The widest possible participation has to be a major objective. Significant countries must be in, not out.

We should be wary of pressures for rich industrial countries to go it alone without taking account of the aspirations of the rest of the world for the kind of consumer goods and services that we take for granted and which at present are heavily dependent on CFCs: not just aerosols and packaging, but refrigerators, air conditioning, insulation, dry cleaning, electronics — even the cushioning in furniture. If, for example, the 2 billion people of India and China were to use refrigeration and air conditioning with CFCs the way they do in New York, the beneficial effect of action in other countries to protect the ozone layer would be seriously at risk.

THE BRITISH RESPONSE: ACHIEVING THE TARGET 10 YEARS AHEAD

What has happened since Montreal? In Britain, certainly, there is much greater public awareness of the need to save the fragile layer of stratospheric ozone that screens us from excessive exposure to ultraviolet radiation. This awareness has developed not only in the scientific community and in government, but among ordinary citizens who are changing from uninformed consumers to 'green consumers'.

The Montreal Protocol also gave an important signal to the world's industry of the urgent need to develop substitutes for CFCs and halons or new technologies. In the meantime we can reduce dependency by using less damaging chemicals or by using less of them by straightforward good housekeeping.

The response of British industry has been extremely encouraging. It fully justifies the policy of aiming wherever possible for voluntary action, the self-regulatory approach. The most notable example has, of course, been with aerosols.

The British aerosol industry expects virtually to phase out the use of CFCs as propellants by the end of next year. Aerosols have accounted for as much as 60% of British CFC use. By the end of 1989 our own consumption of CFCs will be reduced by the 50% Montreal target, 10 years ahead of the date required by the Protocol. I pay tribute to our aerosol industry for this outstanding step. I pay tribute to the consumers of this country who have clearly influenced the industry's decision.

In the USA, CFCs were banned for non-essential aerosol use 10 years ago. This was claimed as a major contribution to protecting the ozone layer. In practice, CFCs found other markets, particularly in foams, refrigeration and air conditioning. Consumption per capita in the USA remains as high as, if not slightly higher than, in the European Community. That is why we are concerned to reduce all CFC uses, in all sectors. British and European industry is already taking a lead here.

ICI, the leading European manufacturer of CFCs, have urged that the review of the Protocol should lead to an eventual phase out of CFCs. Last week they announced that they were investing heavily in a new ozone-friendly chemical that will replace CFCs in refrigeration.

ICI and RTZ/ISC (the other British CFC manufacturers) have joined other European, Japanese and American firms to test the toxicity of the leading candidates to replace CFCs. Both our firms are devoting significant financial, scientific and technical resources to the work.

In addition, there are practical measures being taken by industry now to reduce dependency on CFCs by making better and more efficient use of them, and to introduce recycling techniques, for example in the manufacture of polyurethane foams. Industry is looking at the feasibility of recovering CFCs from discarded refrigerators. The electronics industry is seeking new ways, or improvements of old ways, of cleaning printed circuit boards.

This work was given extra emphasis by the Montreal Protocol and the commitment of the Government to the protection of the ozone layer. The urgency will have been redoubled by this year's increasingly clear evidence that CFCs are implicated in global ozone depletion as well as being powerful greenhouse gases, which contribute to the risk of global warming and consequent climate change.

GOING BEYOND MONTREAL: THE SCIENTIFIC EVIDENCE

The Government's willingness to take precautionary action when necessary is exemplified by our support for the Montreal Protocol. Our policy of basing environmental action on sound scientific evidence of need can scarcely have been better demonstrated than in this area.

We set up a Stratospheric Ozone Review Group (SORG) of independent scientists in the UK to examine and evaluate what is known about ozone depletion and to advise us on the weight of evidence and what further steps need to be taken.

SORG's first report, published in 1987, was an important input to the Montreal negotiations. It reflected the growing scientific concern at the time (when actual depletion was not established) that man-made chemicals were contributing to ozone depletion. This report encouraged us to take a leading role in developing and supporting the European Community's negotiating stance in Montreal.

That was not the end of the story. Scarcely had the ink of Montreal dried than new evidence came in from last year's Antarctic work. I commend the major role played by the US National Aeronautics and Space Administration and I am pleased to see that NASA's Dr Watson is here as a speaker. Certainly many of the leading scientific actors are here today.

During 1988 new evidence flooded in. Although some of it was still contradictory, the Government stepped up its re-examination of the need for stricter measures. Committees of both Houses of Parliament called for the Protocol to be strengthened. The House of Lords called for CFC reductions of 85%. The House of Commons called for consumption to be reduced to 15% of 1986 levels. Both meant the same thing.

The Commons report came out on the day we published the executive summary of our own scientists — SORG. This summary stated in clear and convincing terms that there was now compelling evidence of measurable depletion of global ozone, and not just above Antarctica. SORG found that CFCs were almost certainly involved and that, in order to stabilize chlorine in the stratosphere, worldwide emissions would, as a first step, need to be reduced by 85%. We published this summary and asked SORG to expedite completion of the full text of their second report.

As the full SORG report was sent for printing this September the Prime Minister gave an address to the Royal Society. This major speech is rightly hailed as a milestone in expounding the Government's environmental thinking; in particular, the link with man-made chemicals, which not only threaten the ozone layer but also contribute to global warming. In the speech she said:

> We don't know the *full* implications of the [Antarctic] ozone hole nor how it may interact with the greenhouse effect. Nevertheless it was commonsense to support a worldwide agreement in Montreal last year to halve world consumption of chlorofluorocarbons by the end of the century.

THE CHALLENGE AHEAD: CUTTING CFC EMISSIONS FURTHER AND FASTER

The Prime Minister drew attention to the forthcoming SORG report. It was published on 3 October. On publication day the Minister Malcolm Caithness announced that the Government accepted the need for worldwide emissions of CFCs to be reduced by at least 85% and as soon as possible. I would stress *at least* 85% and that by 'as soon as possible' we think the world should be aiming for a date *before* the end of this century.

At the EC Environment Council last week Lord Caithness urged European colleagues to agree that the Montreal Protocol should be strengthened along the lines we are pressing for, not only to protect the ozone layer but because CFCs are powerful greenhouse gases. Molecule for molecule, CFCs are ten thousand times more powerful greenhouse gases than carbon dioxide, which until recently has been the main preoccupation of scientists and politicians interested in climate change. Most member states at the Council echoed our call, but there are one or two that have still to be convinced of the overwhelming scientific evidence. We will work hard to persuade them to catch up with us.

CLIMATE CHANGE: THE SCIENTIFIC CHALLENGE

By 2030 scientists estimate that, without the Montreal Protocol, CFCs would account for about a quarter of any global temperature rise due to strengthening of the greenhouse effect, whilst carbon dioxide would account for half. Even with Montreal as it stands CFCs will contribute some 13%. A strengthened Montreal Protocol will of course produce a significantly beneficial impact.

It is important that replacement substances must not only have a low ozone depletion potential but also are much less powerful as greenhouse gases.

As with ozone depletion the government recognizes the need for concerted international action to combat the threat of climate change. There is still considerable uncertainty, which needs to be reduced before wide-ranging actions are taken. Nevertheless, there are many actions we can endorse now.

In addition to strengthening the Montreal Protocol, realistic energy pricing of fossil fuels as a sound basis for encouraging energy efficiency must be pursued. Non-fossil fuel energy sources have an obvious role to play, and we include nuclear energy in this category. In June this year the Toronto Conference on the Changing Atmosphere stressed the need for revising the nuclear option as a contribution to reducing carbon dioxide emissions. The role of the World's forests in the dynamics of the carbon dioxide cycle must be recognized. We are active, through the Overseas Development Administration, in encouraging proper forest management that makes both economic and environmental good sense.

THE NEED FOR WORLDWIDE COMMITMENT: THE LONDON CONFERENCE

The Secretary of State, Nicholas Ridley, announced last week that he and the Prime Minister are calling a major international conference on the ozone layer in London next March. The Prime Minister will participate; it will be held in association with the United Nations Environment Programme under whose auspices the Montreal Protocol was successfully negotiated.

It will be a scientific and political conference. We will be inviting Ministers from developed and developing countries, from the east and from the south, from world industry and international organizations. Its purpose is to show that what happens to the ozone layer must be a matter of concern to *all* the countries of the world and that all must meet the challenge. The conference will help the world community to meet this challenge by demonstrating that industry has already developed, or soon will develop, new products and processes that will enable all countries (not just the UK) quickly to reduce the use of CFCs and move to a CFC-free world.

We want our London conference to give a political boost to the first meeting of the parties to the Montreal Protocol, planned by UNEP for April next in Finland. The meeting in Finland will begin the formal process of review of the Protocol. We have offered to host the second and crucial meeting of the Parties planned for April 1990, at which we hope the worldwide reductions we are demanding will be agreed.

We cannot delude ourselves that action by the UK or the European Community, or even all the richest countries of the world today, will save the ozone layer. We need to secure a clear commitment from all world governments, including those who have not yet signed the Montreal Protocol. It is not enough to set objectives. Our Conference in London next year will show how they can be achieved in practice.

British scientists will play a full part in the Protocol's scientific review. As part of that process, an international group of scientists, some of whom are present here today, will meet in London later this week at the Department of the Environment. I am pleased to announce that the UK will contribute $50 000 towards the review. On the linked issue of climate change we fully support the work of the inter-governmental panel, which had its first meeting in Geneva earlier this month. We see the Panel as the main vehicle for carrying forward international activity on climate change, and we were delighted by the appointment of Dr John Houghton, Director General of the Meteorological Office, to lead the scientific assessment team that has been set up.

Finally, I will briefly revert to my theme of the Green Consumer Conferences, like this one and the one we will be holding next year. Not only are they important in persuading governments of the need for action: they give greater knowledge and awareness to ordinary people, who can then exercise their own informed consumer choice.

I salute, for example, *The Green Consumer Guide*, with which several of the people organizing this Conference were involved. I appreciated the opportunity of launching it in September. I understand that the *Guide* has already sold over 120 000 copies.

Rather than attempting to attack firms with what might be called less environmental policies, the *Guide* encourages them to do better. It adopts a positive approach of listing environmentally friendly products. Rather than a blacklist, it offers a greenlist.

There is already healthy competition to get into the next edition, which is planned for next year. This is a more effective incentive than Government bans — whether of aerosols or of any other products.

Industry has already responded promptly to consumer pressure and Government encouragement. The UK will continue to give a lead as we move towards a CFC-free world. We owe it to our children and our children's children.

Part 2
Ozone Depletion

3
The Beginnings of a Problem

IVAR S. A. ISAKSEN
Institute of Geophysics, University of Oslo, Norway

ABSTRACT

Data from the Antarctic have been the subject of much comment in the last few years and offer undisputable evidence that chlorine compounds released by the dissociation of CFCs are affecting the Earth's protective ozone layer. In addition, analysis of ground-based observations indicate that ozone decreases over northern latitudes during the winter. These decreases are not confined to the Arctic region, but can be observed also at mid-latitudes.

Long-term modelling of future ozone changes are highly sensitive to the adopted scenarios for methane, nitrous oxide and CFCs. Nevertheless, over the next decade or so, chlorine levels in the stratosphere will be dependent mainly on previous CFC releases, because of their long chemical lifetime. It is therefore clear that any control measures to limit future chlorine growth in the stratosphere will have minimal impact on stratospheric ozone levels before the turn of the century.

Since 1971, when both Crutzen[1] and Johnston[2] showed that a large fleet of supersonic aircraft (SST) flying in the lower stratosphere could lead to substantial decreases in the Earth's protective ozone layer, large efforts have been made to understand stratospheric ozone chemistry and transport and to predict future ozone changes. In 1974, Molina and Rowland[3] published their first article predicting ozone depletion from human releases of CFCs, and a large part of the research effort has since been concentrated on the fate and influence of these compounds and their oxidation products in the atmosphere. Several outstanding questions have been raised during the past 15 years, including: How stable are the primary chlorine compounds in the atmosphere? What are the main oxidation products in the stratosphere? How efficiently do chlorine compounds break down ozone? Is there already an effect on ozone that can be measured?

Intensive research has made it possible to answer most of these questions. A large-scale programme aimed at studying CFC behaviour in the atmosphere has revealed that the CFCs are practically inactive in the troposphere, making stratospheric oxidation (photodissociation and reaction with excited state atomic oxygen, $O(^1D)$) the only effective loss process. This means that the CFCs have extremely long lifetimes (50–100 years), allowing for transport to the stratosphere. The CFCs are therefore a major source of chlorine in the stratosphere. Measurements in the stratosphere have shown that the inorganic chlorine level has increased during the last few years in agreement with the growth of CFCs and other long lived-primary chlorine species. In addition, the chlorine oxidation cycle of ozone has been demonstrated in the laboratory.

The second question is to know whether there is evidence of ozone depletion. It has now been shown beyond any reasonable doubt that human releases of chlorinated compounds (i.e. CFCs) cause the ozone decrease over the Antarctic that has been observed in recent years, and is reported in the Ozone Trend Report from NASA.[4] There are also indications that chlorine-induced changes in northern hemispheric stratospheric chemistry are occurring. This could lead to substantial changes in future levels of ozone. Hence, it is extremely important to understand the chemical and dynamical processes responsible for ozone distribution in the stratosphere and to take measures to reduce unwanted ozone changes in the future.

OZONE CHEMISTRY IN THE STRATOSPHERE

Ozone is known to be formed in the stratosphere through photochemical dissociation of molecular oxygen at wavelengths shorter than 242 nm:

$$O_2 + h\nu \rightarrow O + O \quad (1)$$

The oxygen atom will then recombine with an oxygen molecule to form ozone:

$$O + O_2 + M \rightarrow O_3 + M \quad (2)$$

(M represents an air molecule.)

Ozone may also photodissociate to form atomic oxygen:

$$O_3 + h\nu \rightarrow O + O \quad (3)$$

Reactions 2 and 3 establish an equilibrium between oxygen atoms and ozone molecules in the atmosphere. Below approximately 50 km, where reaction 2 becomes very fast, ozone is the dominant form of odd oxygen (O or O_3).

Note that, since equilibrium is rapidly established between oxygen atoms and ozone molecules in the stratosphere, loss of either will lead to a reduction in ozone levels.

Ozone may be lost via atomic oxygen and ozone recombination:

$$O + O_3 \rightarrow 2O_2 \qquad (4)$$

The most efficient loss of ozone in the stratosphere is, however, through catalytic cycles of the form:

$$OX + O \rightarrow O_2 + X \qquad (5)$$

$$O_3 + X \rightarrow O_2 + OX \qquad (6)$$

Giving the net reaction:

$$O + O_3 \rightarrow 2O_2$$

The importance of this cycle is that it destroys two odd oxygens without affecting the catalytic compound X, which can be a free radical species in the nitrogen, hydrogen, chlorine and bromine families, such as OH, NO, Cl or Br. For example, theoretical studies indicate that chlorine compounds can be extremely efficient in converting ozone to molecular oxygen. Up to 10 000–100 000 odd oxygen molecules may be lost for each chlorine atom or chlorine monoxide molecule released into the stratosphere. Fortunately, chlorine in this form is not easily transported from sources in the troposphere to the stratosphere. Precipitation (e.g. washout by rainfall) removes most chlorine compounds from the troposphere. The same is true for nitrogen and bromine compounds. To reach the stratosphere, chlorine, nitrogen and bromine have to be tied up in organic compounds that are long-lived and unaffected by precipitation processes in the troposphere. Compounds fulfilling these conditions are the fully halogenated chlorocarbons (F 11, F 12, F 113), nitrous oxide (N_2O) and the halons (H 1211, H1301, H2401). Their fate, concentrations and trends in the atmosphere are therefore crucial to future changes in the stratospheric ozone layer.

The formation of reservoir species (e.g. HCl, $ClONO_2$, HOCl, HNO_3, N_2O_5, HO_2NO_2, HBr) is central to the problem of stratospheric ozone depletion (see Figure 1). When these species are formed, ozone depletion through reactions 5 and 6 is reduced, as most of the reactive compounds are in chemical states that do not contribute to ozone destruction. The formation of reservoir species is particularly important in the lower stratosphere where HNO_3, HCl and HBr are

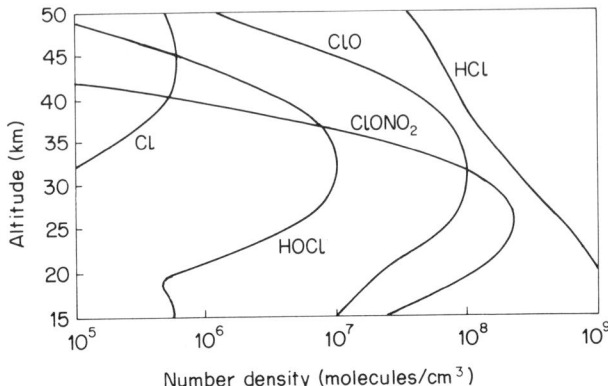

Figure 1 Estimated distribution (1980) of chlorine species in the stratosphere.

the dominant odd species of nitrogen, chlorine and bromine respectively. Figure 1 gives model calculated distributions of chlorine species in the stratosphere. HCl and ClONO$_2$ are clearly dominant in the lower stratosphere.

The relative importance of the different ozone-destroying catalytic cycles is shown in Figure 2. The pure oxygen reactions are seen to be of some importance in the upper stratosphere; lower down, this reaction is unimportant. The nitrogen cycle is the most important ozone-depleting process in the stratosphere, particularly in the middle and lower part of the stratosphere. Hydrogen reactions play a role in the upper and lower parts of the stratosphere. The chlorine cycle has its maximum impact in the upper stratosphere, and is increasing in importance because of rising levels of chlorine in the stratosphere.[5] It should be noted that chemistry plays a more significant role in the upper stratosphere than in the lower stratosphere. To a large extent, transport determines the ozone distribution in the lower stratosphere.

The formation of chlorine nitrate (ClONO$_2$) through the reaction:

$$ClO + NO_2 + M \rightarrow ClONO_2 + M \qquad (7)$$

represents an important link between the active chlorine and nitrogen cycles in the stratosphere, as ClONO$_2$ ties up both chlorine and nitrogen in an inactive form with regard to ozone depletion. This means that, the more chlorine levels increase in the atmosphere, the less efficient the nitrogen cycle becomes (reactions 5 and 6). Consequently the chemistry of ozone destruction is far from linear. The rate of ozone destruction and future ozone levels are therefore dependent on the combined growth of chlorine and nitrogen source gases in the stratosphere (CFCs and N$_2$O).

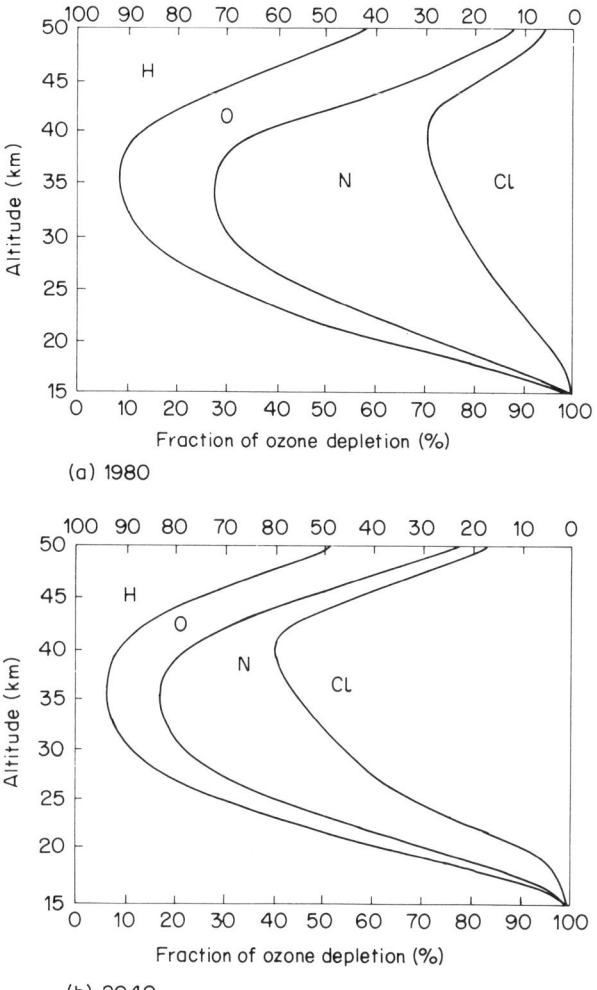

Figure 2 Fractional depletion of ozone from chemical reactions with hydrogen, oxygen, nitrogen and chlorine in the years 1980 and 2040 (percentage of total loss at each height). (Note that the depletion scales can be read in both directions.)

INCREASES IN STRATOSPHERIC CHLORINE

Inorganic chlorine in the stratosphere (Cl_y = Cl + ClO + HCl + $ClONO_2$ + HOCl) derives from several organic sources. With the exception of occasional inputs from volcanic eruptions, the only natural source gas of importance is CH_3Cl, which contributes approximately 0.6 ppb to stratospheric chlorine, and possibly CCl_4, which is partly of natural origin. CCl_4 contributes approximately

0.4 ppb to stratospheric Cl_y. The present level of Cl_y is approximately 3 ppb. This means that already man-made chlorine releases account for 70–80% of the chlorine load in the stratosphere. Future chlorine levels are predicted to increase further as a result of man-made releases, regardless of control measures, with the bulk of chlorine coming from F 11 and F 12, whose atmospheric concentrations are continuing to increase at more than 5% per year (Table 1) even though the release rates have remained practically constant since 1974 when the first warnings about CFCs were given.[3] This phenomenon reflects the stability of the fully halogenated CFCs in the atmosphere. The prolonged residence time of CFCs 11 and 12 of 50–100 years means that changes in release rates are not reflected in atmospheric concentrations for a decade or more. Thus any action to reduce the unwanted effects on stratospheric ozone has to be taken well in advance of possible changes.

Table 1 Ozone depletion potential for the different chlorine and halon species

Compound	Lifetime	ODP
CH_3CCl_3	6.2	0.13
HCFC 22	17.0	0.038
CFC 11	57.7	1.0
CFC 12	96.4	0.86
CCl_4	50.8	1.19
CFC 113	103.0	0.93
CFC 114	247.0	0.82
CFC 115	548.0	0.40
H 1211	17.7	3.94
H 1301	72.1	9.82
H 2402	23.3	6.55

The comparison is made for the same mass released of each compound with the efficiency of CFC 11 set equal to 1

OZONE DISTRIBUTION AND CHANGES

Since our main concern with ozone layer depletion is the ability of the ozone layer to filter out harmful ultraviolet radiation from the Sun, it is natural to examine ozone distribution with regard to latitude and at different times of the year. Model studies are aimed at predicting how total column ozone will change in the future, and Figure 3 shows total column ozone as a function of month and latitude: ozone levels vary with both. Maximum values are obtained during late winter and spring at high latitude, with low values during summer at mid- and high latitudes, and lower values at low latitudes.

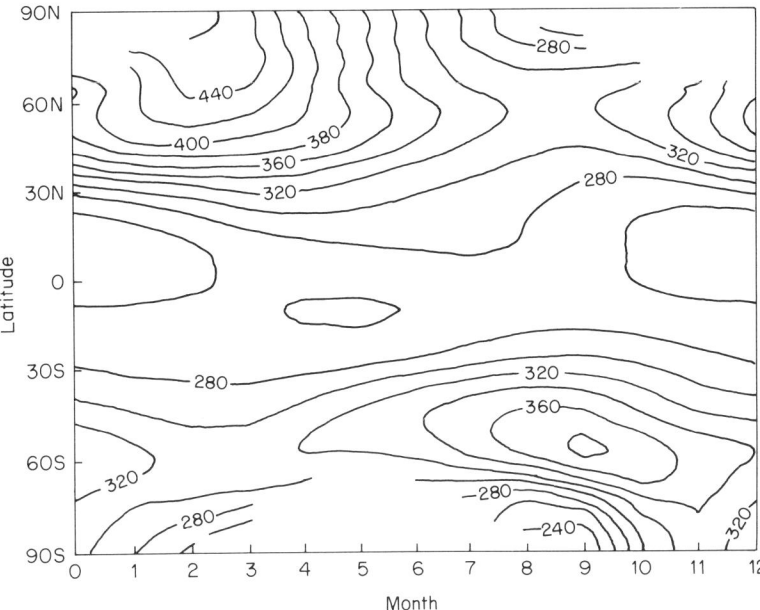

Figure 3 Total observed ozone column densities, 1979–86, as a function of latitude and season. The values are given in Dobson units (1 Dobson unit = 2.7×10^{16} molecules/cm^2 at normal temperature and pressure).

Ozone distribution results from the combined effects of chemistry and transport. Ozone is produced in the middle and upper stratosphere at low latitude, and efficiently transported to high latitudes where it is brought down to lower levels. The result of this process is increasing concentrations, which is most pronounced during the winter months.

Figure 3 shows average long-term values. It is important to recognize, however, that the ozone column may vary substantially from day to day and along a latitudinal circle, owing to dynamic processes.

The most extensive and continuous measurements of total column ozone come from Arosa in Switzerland, where the measurements date back almost 60 years (Figure 4). The column varies substantially from year to year, making trends in total column ozone more difficult to detect. However, during the last decade or two, there are clear indications that total ozone has decreased.

The analysis by the Ozone Trend Panel[4] involved a large number of stations in the Northern Hemisphere, and indicates that winter ozone has declined substantially since the beginning of the 1970s, maybe by as much as 3–5%, which is a substantially larger ozone loss than predicted by the models (Figure 5).

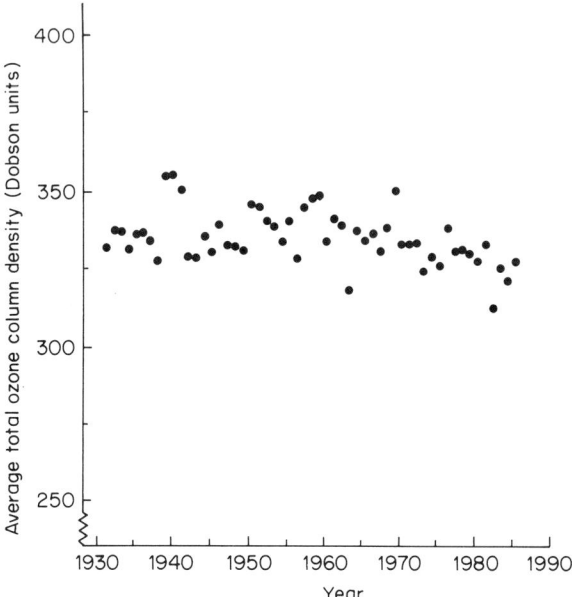

Figure 4 Annual average total ozone column density over Arosa, Switzerland in Dobson units.

The Ozone Trend Panel also examined changes in the vertical distribution of ozone in the stratosphere, and made comparisons between observational data and model studies. The results indicate that there is a detectable ozone decrease in the upper stratospheric region after 1980 due to CFC releases. Caution is warranted, however, as part of the decrease is probably due to the 11-year solar cycle effect (see Chapter 5). This is estimated to account for a 1 or 2% decrease in total column ozone between the solar maximum (e.g. 1979) and the solar minimum (e.g. 1985). The change might be somewhat larger locally in the 40 km region. Discrepancies between the observational data and the model calculation were not found in the upper and middle stratosphere.

If models underestimate winter ozone changes, as indicated by the ozone trends study, the discrepancy probably relates to the lower stratosphere. One distinct possibility is that there are Arctic polar sinks for ozone analogous to but less pronounced than the heterogenous reactions in Antarctica connected to the formation of polar stratospheric clouds.

The occurrence of the Antarctic ozone hole, which has been clearly demonstrated by ground-based satellite observations over the past 10 years, and which has become particularly pronounced during the last few years, is undoubtably the result of increasing chlorine levels in the stratosphere. This has

THE BEGINNINGS OF A PROBLEM

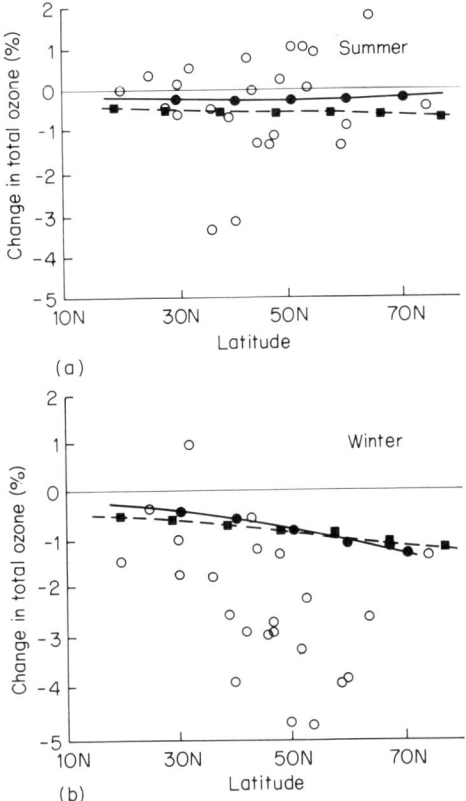

Figure 5 Calculated changes in total ozone at mid- and high latitude northern hemispheric stations over two consecutive 11-year solar cycles (1965–75 to 1976–86), and model-calculated changes. ○, observations; ●, Oslo model calculations; ■, AER model calculations.

been established through intensified research during 1986–87 as reported by the Ozone Trend Panel[4] (see Chapter 5). It is probable that this man-made phenomenon of the ozone hole will prevail for decades unless meteorological patterns change substantially.

ESTIMATES OF FUTURE OZONE DEPLETIONS

Future ozone changes will depend strongly on how CFC release rates are controlled. Figure 6 shows the results from three combined scenarios involving methane, nitrous oxide and CFCs.[6] They are based on continuous growth in the release rates without any control measures, release rates based on Montreal scenarios, and assumptions that all fully hologenerated CFCs in the Montreal

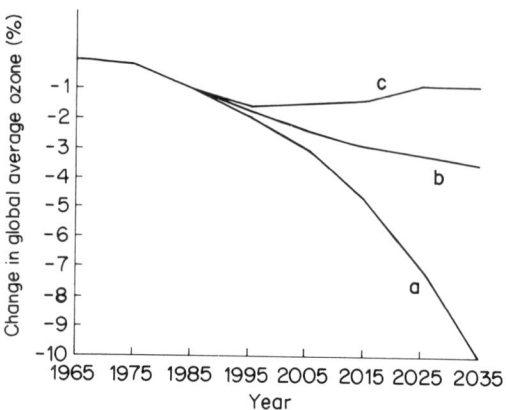

Figure 6 Global average ozone changes with time for three selected scenarios: **a**, continuous growth scenario; **b**, Montreal protocol scenario; **c**, scenario with all fully halogenated CFCs replaced by HCFC 22.

scenario are replaced by HCFC 22, a CFC with less ozone depleting potential (ODP) than CFC 11 or CFC 12. It can be seen that unlimited growth gives unacceptable ozone depletions, and even the scenario based on the Montreal protocol gives substantial ozone depletion during the next decades. The use of the alternative trace gas HCFC 22, on the other hand, will reduce the amount of ozone depletion. The percentage changes in Figure 6 refer to global average changes. Table 1 shows the chemical lifetimes and ODPs for some of the main CFCs and some alternatives (HCFCs) as well as the most common halons.

In general, the ODP is substantially less for compounds that contain hydrogen atoms. It is also clear that halons are extremely efficient ozone depletors.

CONCLUSIONS

Chlorine compounds act as ozone sinks in the stratosphere through catalytic chemical reactions involving chlorine monoxide (ClO), and atomic chlorine (Cl). Measurements show that the levels of chlorine in the stratosphere have increased over the last few years. As much as 70–80% of stratospheric chlorine is the result of man-made releases. There is increasing evidence that the enhanced chlorine levels are affecting stratospheric ozone to the extent that ozone reductions can now be measured. Ozone depletion over the Antarctic is clearly the result of increased chlorine levels in the stratosphere. Analysis of ozone data in the Northern Hemisphere strongly indicates that there has been a decrease in winter ozone levels over the past two decades, which is unlikely to be due to natural variations. Very recent measurements of a strongly disturbed Arctic stratosphere in the winter months strengthens the suspicion that man-made chlorine, and possibly bromine, compounds are responsible for the observed decrease.

Atmospheric chemical models used to simulate future changes in the ozone layer predict that chlorine levels will continue to increase over the next few decades, and this will lead to substantial ozone depletion unless CFC releases are curtailed. One way to reduce the impact of chlorinated compounds on the stratospheric ozone layer is to use alternative compounds with very low ozone depletion potential. Another way is to develop technologies that do not require CFCs at all.

The fact that a clear link has now been established between the Antarctic ozone hole and man-made chemicals, and that the Arctic stratosphere is more disturbed than previously believed, makes it necessary to monitor stratospheric ozone very closely. It is also necessary to take strong measures to limit the growth of stratospheric chlorine.

REFERENCES

1. P.J. Crutzen, Ozone production rates in an oxygen, hydrogen, nitrogen-oxide atmosphere. *J. Geophys. Res.,* **76,** 7311–7327.
2. H.S. Johnston, Reduction of stratospheric ozone by nitrogen oxide catalysts from supersonic transport exhaust. *Science,* **173,** 517–522 (1971).
3. M.J. Molina and F.S. Rowland, Stratospheric sink for chlorofluoromethanes: chlorine atom catalyzed destruction of ozone. *Nature,* **249,** 810–814 (1974).
4. NASA, *The Ozone Trend Panel.* (1989), in press.
5. I.S.A. Isaksen and F. Stordal, Ozone perturbations by enhanced levels of CFCs, N_2O and CH_4: a two-dimensional diabatic circulation study including uncertainty estimates. *J. Geophys. Res.,* **91,** 5249–5263 (1986).
6. B. Rognerud, I.S.A. Isaksen and F. Stordal, Model studies of stratospheric ozone depletion. (1989), in press.

4
Numerical Modelling of Ozone Perturbations

RICHARD S. ECKMAN AND JOHN A. PYLE
University of Cambridge, UK

ABSTRACT

Problems in modelling stratospheric ozone perturbations are discussed. Model formulation is considered first. The choice of formulation (dimensionality, degree of feedback, etc.) introduces approximations and limitations to the model. Other uncertainties include the kinetic data, the completeness of the model chemistry and the reliability of the various source gas emission rates. Results from a particular two-dimensional model are presented, employing a variety of scenarios for emissions of CFCs and other gases. All the calculations presented show global ozone depletions in the next century, but the degree of depletion depends critically on the different emission rates. In particular, the importance of methane, the increase in which can partially compensate the effect of CFCs, is highlighted.

INTRODUCTION

The present level of interest in ozone depletion can be traced back to the theoretical papers by Crutzen[1] and Johnston,[2] who suggested that oxides of nitrogen might reduce stratospheric ozone concentrations, and of Stolarski and Cicerone[3] and Molina and Rowland,[4] who first proposed that chlorine compounds could destroy ozone in the stratosphere. The observational study of Farman *et al*.[5] showed dramatically that ozone in Antarctica *was* being destroyed, and recent campaigns[6,7] have demonstrated beyond reasonable doubt that the destruction is due to chlorine compounds, which arise mainly from destruction of CFCs. Theoretical, or numerical, models have played an equally

---

Ozone Depletion: Health and Environmental Consequences
Edited by R. Russell Jones and T. Wigley
© 1989 John Wiley & Sons Ltd

important role in stimulating both scientific and public awareness of stratospheric pollution, and will continue to be extremely important in discussion on the strengthening of the Montreal Protocol.

This chapter describes in simple terms the elements of numerical models of the chemistry of the atmosphere. Some calculations using a two-dimensional model will be described. An important point to realize is that models contain many approximations and hence their predictions must be treated with appropriate caution. We describe some of the main areas of uncertainty, demonstrating them in some cases by further calculations.

NUMERICAL MODELLING OF THE ATMOSPHERE

The atmosphere is a fluid of great physical and chemical complexity, and analytic solutions to all but the most simple (although very often important) problems are impossible. To solve problems relating to the long-term evolution of the atmosphere necessarily calls for the use of a numerical model, which can often be integrated only on the largest computers available. The numerical model will comprise a set of equations that describe the dynamical meteorology of the atmosphere, the transfer of heat and radiation within the atmosphere, the chemical interactions taking place, and any physical processes (e.g. cloud formation) that may be relevant. In addition, surface processes, both natural and anthropogenic, must also be described in terms of equations.

For solution of the equations, the atmosphere is generally divided into a grid or lattice. The size of the grid boxes, and hence the spatial resolution, will vary from model to model but would typically be a few (approx 2–4) kilometres vertically and a few hundred kilometres horizontally.

The system is highly coupled. For example, ozone is not only chemically active but its distribution affects radiative transfer and, thus, atmospheric heating and the associated wind motions. Physical processes are often critically temperature dependent. Because of this complexity, in all approaches to modelling the atmosphere some simplifications must be made. A traditional approximation has been to reduce the number of spatial dimensions in the problem. Early chemical models considered the atmosphere as a function of altitude only; these are one-dimensional models. Two-dimensional models, which describe the atmosphere in terms of altitude and latitude, have been used for perhaps 15 years, and extensively in the last few years. These models can include very detailed chemistry, as well as providing a reasonable description of some aspects of atmospheric transport. There is, however, an important component of transport in two-dimensions (that associated with large scale atmospheric waves) that cannot be calculated *a priori* in a two-dimensional model. This transport must be included, and this is usually done by appeal to some kind of eddy diffusivity. This area of two-dimensional modelling has been one of active research in recent years,[8] and significant progress has been made to place the models on a firmer physical basis.

Three-dimensional models, in which latitude, longitude and altitude are all treated, have as yet been little used for chemical perturbation studies. They demand major computing resources, which are only just becoming available. Nevertheless, there are now a number of groups active in developing and running three-dimensional models including chemistry and some impressive results are beginning to appear.[9] During the next few years, efforts to use these models for perturbation studies will certainly increase.

SOME UNCERTAINTIES

There are a number of important uncertainties that must be recognized when making predictions of ozone change into the next century. In this section some of these uncertainties will be introduced; specific examples of sensitivity studies that demonstrate these uncertainties are presented in the next section.

Model formulation

A question of central importance is clearly the extent to which models can be used as predictive tools. How is the predictive capability affected by model formulation? Evidently, the most detailed models available, including all the myriad chemical and physical interactions, should be used. Even these will not be perfect, as we can appreciate by analogy with the most sophisticated weather forecasting models. The 'skill' in these models is impressively high, but they have their limitations.

What, then, should be said about models that are less detailed, for example, the two-dimensional and one-dimensional models referred to earlier? Clearly such models must be used within their obvious limitations. One-dimensional models provide an excellent framework in which to study chemical interactions. Their treatment of atmosphere transport is rudimentary and hence, at best, they provide only an indication of possible global changes. Two-dimensional models do include, albeit in an approximate fashion, a description of vertical and horizontal transport. These models are therefore indicative of likely effects as a function of latitude and altitude. A prerequisite of one- and two-dimensional models is that they represent the present atmosphere satisfactorily. They must therefore be compared against available data before they are used for predictions. A satisfactory comparison does not prove the model's predictive capability, but an unsatisfactory one would evidently cast doubt on the model.

From the foregoing it should be clear that there is no quantitative index that can be used to assess the suitability of any particular model; rather, the assessment is subjective and must be based on a cautious examination of the underlying model formulation and its ability to reproduce the present atmosphere as well as the historical trends in trace gases.

There are many different types of two-dimensional model. Some models reduce their complexity by ignoring certain feedback or coupling processes. For example, some models use the calculated ozone to calculate heating rates, which can be employed in the calculation of the temperature and wind fields. Other models simply specify the temperature and wind fields. This latter procedure can have some advantages in that the temperature, in particular, can be specified accurately and the wind fields can be specified to produce good tracer distributions. On the other hand, it may have serious disadvantages. In particular, the important feedback between ozone and temperature is lost and this could significantly alter the details of predicted ozone change (see the next section).

Emission data

Naturally, the magnitude of any predicted effect of the CFCs on ozone will depend on the quantity of CFCs emitted into the atmosphere. Good emission data are needed. Future emissions depend on a host of socioeconomic and political factors and can be difficult to predict. Most important, though, is the fact that ozone is affected by gases other than the CFCs, which are themselves changing. Thus predictions will also depend on the emission rates assumed for the halons (bromine-containing source gases), methane, nitrous oxide and carbon dioxide. As shown in the next section, increases in methane lead to increases in ozone, thus compensating to some extent the ozone depletion due to CFCs. The net effect on ozone depends sensitively on the emission rates for methane. The present emission of and trends in methane are not completely understood; extrapolating the trend into the next century is very uncertain.

Missing or inadequately understood chemistry

Calculations of ozone depletion depend critically on the inclusion of an adequate kinetic database in the models. Figure 1 provides a striking demonstration of the way in which predictions with one particular model have varied as a function of the time the calculation was performed. The many fluctuations in the results were caused mainly by new laboratory rate data being included in the models. Although our knowledge of the relevant reaction rates has improved substantially during the past 15 years, this cannot preclude the possibility of further changes, which would subsequently alter the ozone predictions.

Another reason for the fluctuations in Figure 1 was the inclusion at various times of chemical processes which had hitherto been ignored. For example, the realization that the reaction

$$ClO + NO_2 + M \rightarrow ClONO_2 + M$$

could tie up 'active' chlorine and nitrogen ('active' meaning that these compounds could directly destroy ozone) had a major impact on predictions.

More recently, the unravelling of the mysteries of Antarctic chemistry has shown that, in that region, a number of processes that had previously not been included in models are important. These include the chemistry of the chlorine monoxide dimer $(ClO)_2$, interactions between ClO and BrO, and a number of evidently very important reactions on the surfaces of ice crystals in polar stratospheric clouds (PSCs).

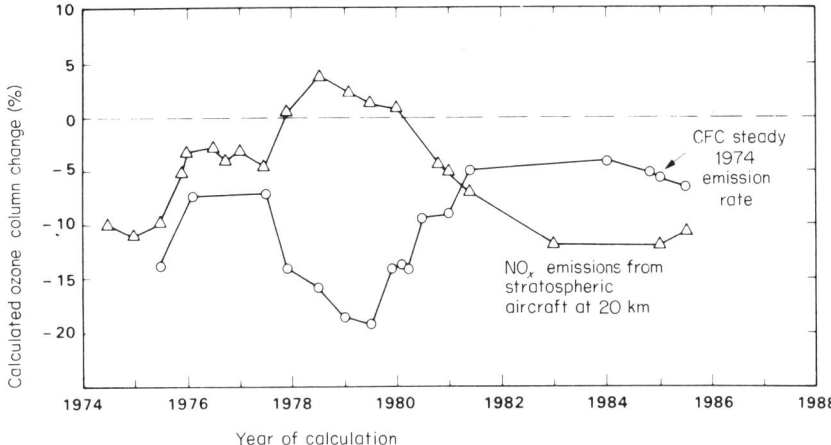

Figure 1 Steady-state column ozone depletion calculated with the Lawrence Livermore National Laboratory one-dimensional model for two assumed perturbations: △, nitric oxide emissions from a hypothetical, large fleet of supersonic aircraft at 20 km; ○, CFC 11 and CFC 12 emissions at the 1974 rate. The calculations, over an 11-year period, reflect the impact in the model of inclusion of new reactions and changing kinetic and photochemical data, transport parameterizations and boundary conditions. (Source: WMO.[8])

The surface reactions are particularly interesting. Molecule–molecule reactions, which are very slow in the gas phase, will proceed on a much shorter time scale on ice surfaces. For example, it is now thought that in the Antarctic lower stratosphere, the surface reaction

$$ClONO_2 + HCl \rightarrow Cl_2 + HNO_3$$

produces chlorine molecules, which can be rapidly photolysed to initiate ozone destruction cycles. The importance of the PSCs in Antarctica raises the question of the global impact of this kind of reaction. In particular, if such reactions also occur on the sulphate aerosol present throughout the lower stratosphere, present estimates of ozone depletion would almost certainly be underestimates. Two observations point to the possible global impact of PSCs. First, the recent report of the Ozone Trend Panel has shown that the observed ozone depletion in the Northern Hemisphere north of 55°N in winter, is considerably larger than models predict. Secondly, Brune et al.[10] have observed concentrations of ClO in the high-latitude lower stratosphere that are higher than model predictions.

The lower stratosphere

In making predictions of ozone change, the lower stratosphere is the most important area to be understood. The bulk of the ozone layer lies between the tropopause and about 30 km. Changes to ozone in this region therefore have the largest impact on the variation of the ozone column, and hence, for example, on the penetration of solar ultraviolet radiation. Although (see later) the largest *percentage* reductions are predicted for the upper stratosphere around 40 km, the ozone density is low here and these changes do not have a corresponding influence on the percentage reduction of the ozone column.

Unfortunately, the lower stratosphere is a difficult area to model. The chemical and dynamic time scales for ozone are comparable here, and hence both processes must be understood well. Small errors in either could have a large impact on the predicted ozone distribution. The radiative balance is also difficult to model. The net radiative heating results from the balance of many comparably small terms. Errors in modelling the radiative balance, which would then affect the modelled transport, are quite likely. Eckman et al.[11] have argued that small changes in temperature due to changing concentrations of carbon dioxide can alter the ozone perturbation by a large amount. The chemical balance is also interesting in this region. The effect of increasing chlorine levels is to reduce ozone in percentage terms, principally in the upper atmosphere and to a lesser extent in the lower stratosphere. Increases in methane are expected to increase ozone, by the largest percentage amounts in the troposphere and to a lesser extent in the lower stratosphere. In the lower stratosphere the two processes are competing and the net effect depends crucially on the emission rates of the chlorine source gases and of methane.

Evidently, the behaviour of the lower stratosphere is central to an understanding of future ozone change. It is a difficult area to understand, with many competing processes. The modelling of these processes themselves, as shown above, is often subject to considerable uncertainty. A sustained effort to improve our understanding of the region, by both modelling and observational studies, is called for.

CALCULATIONS WITH A TWO-DIMENSIONAL MODEL

In this section, a number of two-dimensional model simulations using different emission scenarios will be presented. The model used is the Cambridge University two-dimensional model, which has been described extensively in the literature.[11-14] The model calculates the zonal means of wind, temperature and a variety of chemical constituents. The model domain extends from pole to pole

NUMERICAL MODELLING OF OZONE PERTURBATIONS 33

and from the ground to approximately 90 km, although in these calculations the chemistry is treated up to only about 60 km. The model uses a 4 h time step and the vertical and horizontal resolutions are 0.5 in $\ln(p/p_0)$ (approx. 3.5 km) and $\pi/19$ (approx. 1000 km) respectively.

A number of different scenarios have been run. These are detailed in Table 1. In all the runs, common assumptions were made regarding the increases of carbon dioxide, nitrous oxide, HCFC 22, carbon tetrachloride, methylchloroform and the halons. Four different assumptions were made about the other CFC emissions (taken here to include just CFC 11, CFC 12 and CFC 113). CFC 11 and CFC 12 are the major chlorine carriers in these runs. CFC 113 is assumed to be representative of all other CFCs, and its flux was adjusted accordingly. All assumed a total emission for these gases of 1095 kt in 1985. In run CU5, constant emission was assumed after 1985. In CU4, a 3% per annum growth rate in emission of the CFCs was assumed. CU7 assumed emissions based on the Montreal Protocol, with a 50% reduction in the emissions compared with 1985 by the end of the century. In these three runs, methane was taken to increase at 1% per annum. Run CU6 examined sensitivity to the methane emissions. In this run the surface mixing ratio of methane was held fixed at its 1985 value and constant fluxes of the CFCs at the 1985 level were used. CU6 should be compared with CU5. The emissions of the CFCs and of carbon dioxide in these cases are somewhat higher than have been used in some calculations. For a calculation with this model using lower CFC emission rates, see SORG[15].

Table 1 Trace gas scenarios used in the Cambridge University two-dimensional model

Gas	1985 emission/concentration	Scenario number	Annual increase (%)
1. Common trace gas assumptions			
Carbon dioxide	345 ppmv		0.76
Nitrous oxide	300 ppbv		0.25
Methane	1.7 ppmv		1.0 (0 for CU6)
HCFC 22	86.9 kt/year		3.0
Carbon tetrachloride	76.2 kt/year		3.0
Methylchloroform	534.5 kt/year		3.0
Methyl chloride	0.62 ppbv		Zero
Halon 1211	8.2 kt/year		Zero
Halon 1301	8.2 kt/year		Zero
2. 'Controlled' CFCs (CFC 11, CFC 12, CFC 113)	Total of 1095 kt/year	CU5	Zero
		CU4	3.0
		CU7	Montreal
		CU6	Zero (fixed CH_4)

All runs started from a 1985 atmosphere, which was produced by running the model with representative fluxes of the various gases, starting in 1950. In general, the 1985 model atmosphere corresponds well with the available observations.

Figure 2 shows the upper stratospheric mixing ratio of active chlorine, the final product following breakdown of the CFCs and other chlorine source gases, as a function of time. With 3% growth in CFC emissions, there is a rapid growth in active chlorine in the stratosphere to perhaps five times present atmospheric levels by the year 2040. Even under the terms of the Montreal Protocol the model shows that active chlorine will double by 2040. A cut back in emissions to approximately 85% of the present values is required to stabilize the atmospheric chlorine mixing ratios (for example see SORG[15]). In run CU7, one half of the active chlorine by 2040 originates from CFC 11 and CFC 12 while carbon tetrachloride (17%), methylchloroform (13%) and CFC 113 (10%) all make reasonable contributions. HCFC 22 contributes only about 3% to the active chlorine levels in the stratosphere in run CU7 in year 2040.

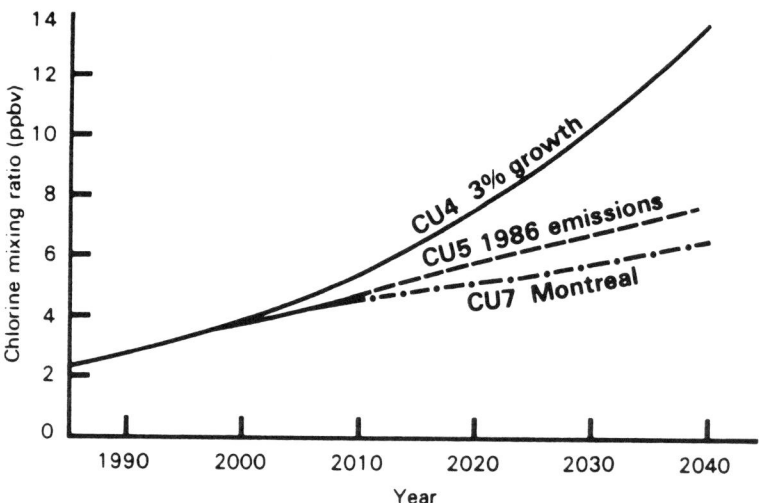

Figure 2 Accumulation of active chlorine in the upper stratosphere as a function of time, predicted by the Cambridge University two-dimensional model. (Based on SORG,[15] reproduced by permission of the Controller of Her Majesty's Stationery Office.)

Figure 3 presents the calculated global ozone changes for these runs, starting in 1985. Ozone in the model is reduced in all the cases presented. For cases with constant emissions (CU5) or considering the Montreal Protocol (CU7), the depletion by 2040 is 1% or less, a reflection of the compensating effects on increasing CFCs, methane and carbon dioxide. Note that if CFC emissions are assumed to increase (CU4) the ozone depletion grows rapidly after about the year

2015. On the other hand, runs (not shown, but see SORG[15]) assuming the Montreal Protocol but which employed lower emission rates for CFCs 11, 12 and 113 calculate a small global ozone increase by 2040. Evidently, the net effect is sensitive to the balance between the competing processes and depends critically on the emission rates used in the model. This is demonstrated further by run CU6, in which methane levels are held fixed at the 1985 values, but with other concentrations following run CU5. The run with constant methane produces significant global ozone depletion which is increasing with time, whereas in CU5 the ozone depletion levels off at around 1%. It should be stressed that the reasons for the present increases in methane are incompletely understood and that any extrapolation of the increase to the middle of the next century must be extremely uncertain.

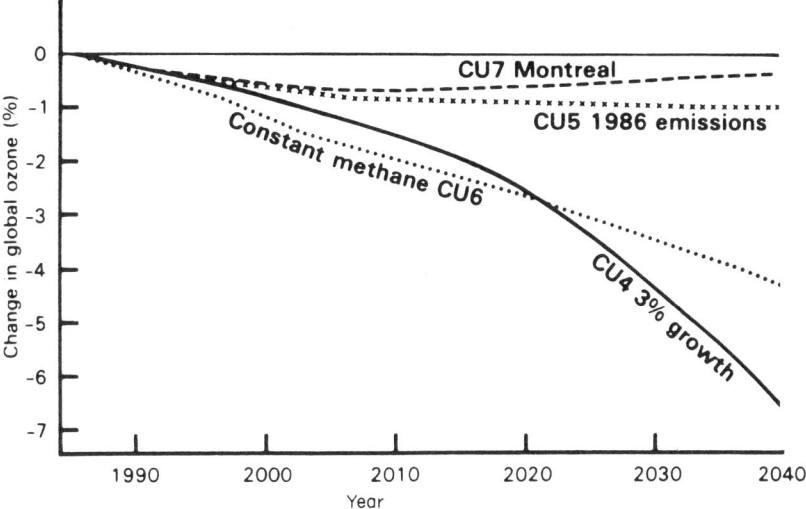

Figure 3 Predicted change in global-mean ozone *versus* time calculated with the Cambridge University two-dimensional model. (Based on SORG,[15] reproduced by permission of the Controller of Her Majesty's Stationery Office.)

Figure 4 shows latitude–time sections of the change in the vertically integrated ozone amount ('column ozone') between 1985 and 2040 for runs CU4 (the 3% growth case) and CU7 (the Montreal Protocol case). The 'footprint' of ozone depletion is very different in the two cases. In run CU4, the largest depletions are predicted in high latitudes in spring with significant depletions at all latitudes. In contrast, the Montreal Protocol run shows very low depletions in low latitudes and even predicts regions in which the column ozone will increase, for example over the Northern Hemisphere pole in summer.

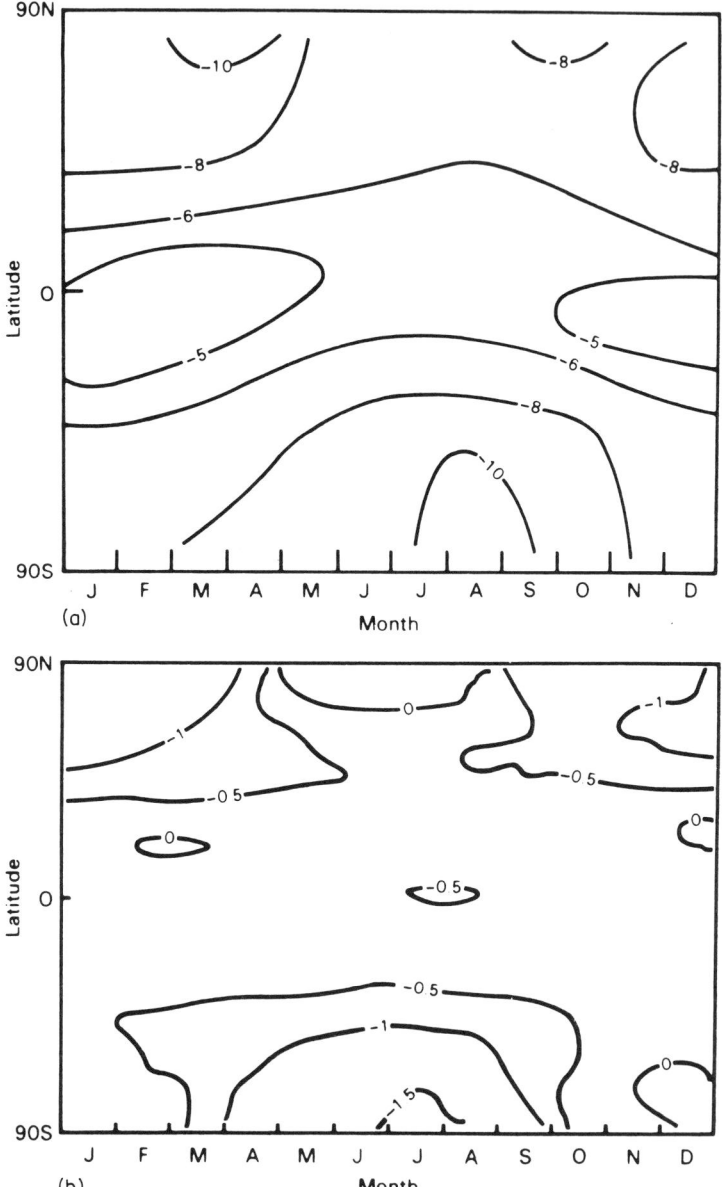

Figure 4 Percentage change in column ozone between 1985 and 2040 as a function of latitude and month for (a) run CU4 (3% growth), and (b) run CU7 (Montreal Protocol). (Reproduced from SORG[15] by permission of the Controller of Her Majesty's Stationery Office.)

Figure 5 shows the latitude *versus* height sections of percentage ozone change between 1985 and 2040 for CU4 and CU7. The patterns are quite similar, with ozone depletion in the upper stratosphere being compensated, to a greater or lesser extent, by ozone production in the lower stratosphere. With a 3% growth in CFC emissions, and approaching 14 ppbv active chlorine, the ozone depletion in the upper stratosphere (Figure 5(a)) is large, being greater than 50% in high latitudes. Even under the Montreal Protocol the model still predicts depletions greater than 20% in parts of the upper stratosphere.

Although the net effect on the ozone column may be small, the local changes are often large. This arises because of the compensating processes discussed earlier. Increasing levels of methane, the increased penetration of ultraviolet radiation consequent on ozone loss in the upper stratosphere, and the cooling of the stratosphere with increasing concentration of carbon dioxide all lead to production of ozone. There is a fine balance between the different influences on stratospheric ozone, which makes it difficult to predict future changes. Note, for example, that in CU7 the ozone concentration has increased around 30 km over the southern pole, leading to the increase in column ozone seen in Figure 4(b). As stressed earlier, the net effect in the lower stratosphere is a critical determinant for global effects. Different models used to assess ozone changes sometimes show quite different lower stratospheric responses (see SORG[15]).

To demonstrate further the sensitivity to emission rates as well as the crucial role of the lower stratosphere, Figure 6 shows the cross-section of percentage ozone depletion from CU6 with surface methane levels held constant. Changes in the upper stratosphere are between those predicted for CU4 and CU7, as might have been expected from the different level of active chlorine in this run (which has the same CFC emissions as CU5, see Figure 2). The lower stratosphere and troposphere exhibit very different behaviour from that seen in CU4 and CU7 (Figure 5). There is only a small region of ozone increase in the tropical upper troposphere and lower stratosphere, and in high latitudes within the troposphere. Not surprisingly, the lower levels of methane oxidation lead to less ozone production and a global depletion considerably larger than in the other two runs.

While the global changes to 2040 predicted in most of the runs are quite small, the local effect, as stated earlier, can be much larger. Ozone controls the heating of the stratosphere principally by absorption of solar radiation and also by radiative transfer in its 9.6 μm band. These factors, and the radiative effects of increasing carbon dioxide, are included in this model, although not in all models used for ozone assessment. Large local ozone changes will alter the temperature structure. Figure 7, which shows the temperature perturbations between 1985 and 2040 for runs CU4 and CU7, should be compared with the ozone changes shown in Figure 5. Large temperature reductions are predicted for the upper stratosphere due to the combined effects of increased carbon dioxide and reduced ozone. The large difference (approx. 8 K) between the two runs in the upper stratosphere is due solely to the ozone difference between the calculations, since

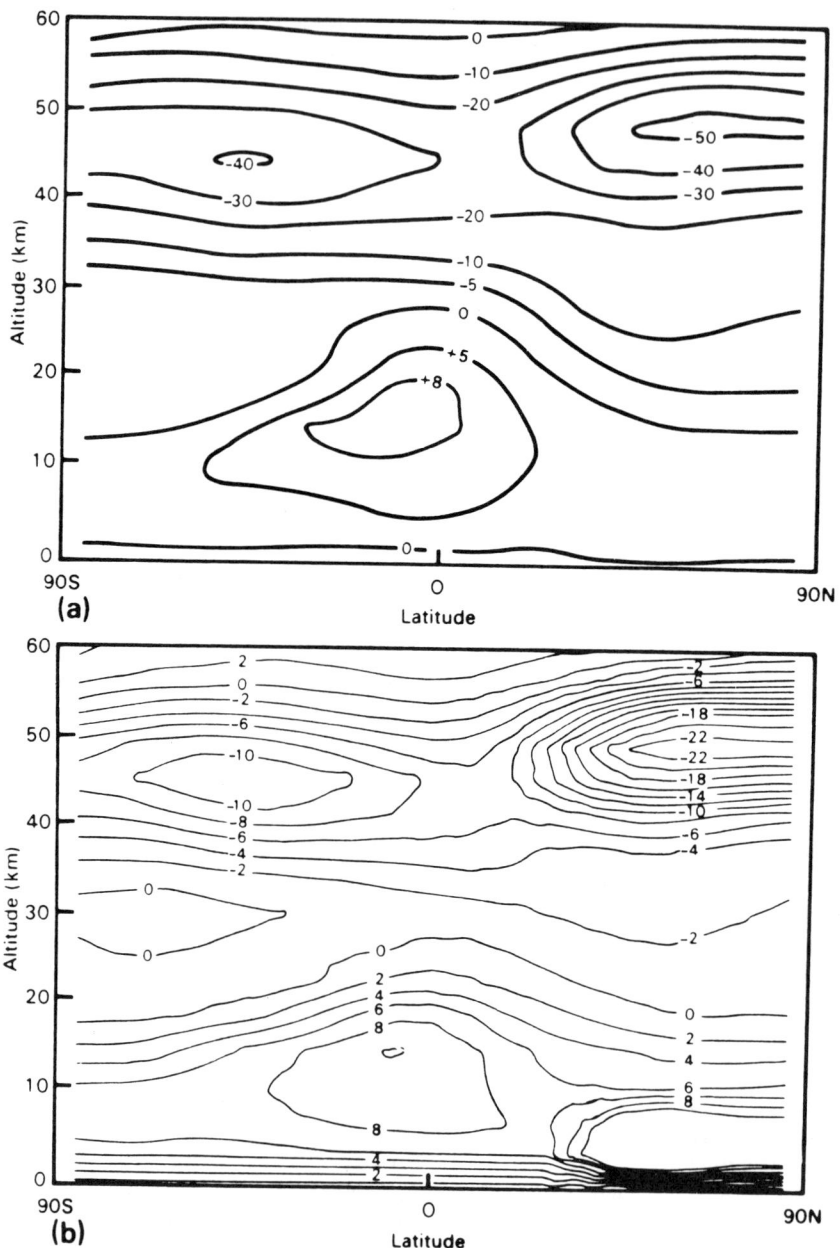

Figure 5 Latitude *versus* height cross-sections of percentage ozone change in December between 1985 and 2040 calculated with the Cambridge University two-dimensional model: (a) run CU4, (b) run CU7. (Reproduced from SORG[15] by permission of the Controller of Her Majesty's Stationery Office.)

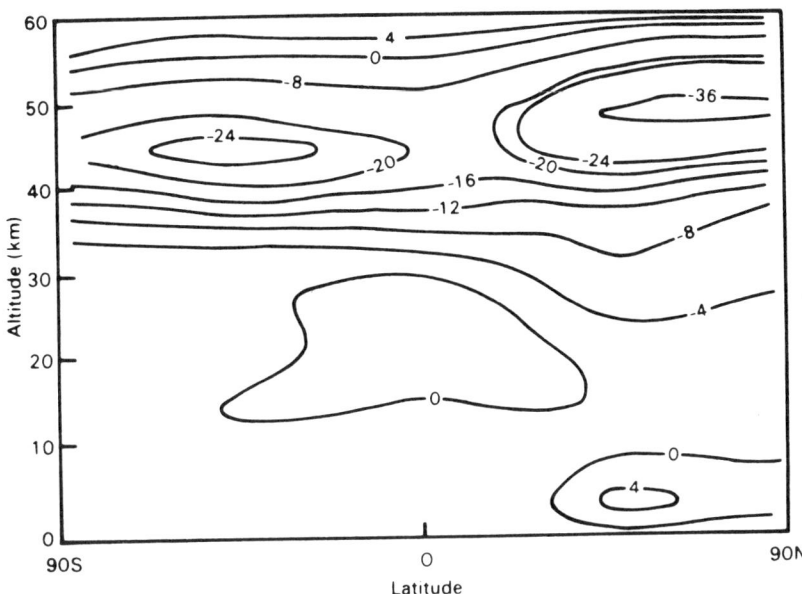

Figure 6 Latitude *versus* height cross-section of percentage ozone change in December between 1985 and 2040 for run CU6 (fixed surface methane concentration). (Reproduced from SORG[15] by permission of the Controller of Her Majesty's Stationery Office.)

both used the same carbon dioxide growth rates. Models that do not include the feedback between ozone and temperature are omitting an extremely important climatological feedback (and possibly as a consequence overestimating the ozone depletion). There are also interesting changes in the lower stratosphere. Eckman et al.[11] have discussed the sensitivity of ozone change to temperature changes in the lower stratosphere.

These temperature changes will be expected to have a significant impact on atmospheric winds and, hence, constituent transport. Present models cannot include these dynamical feedbacks adequately. Their impact needs to be considered in a more complete dynamical model.

CONCLUSIONS

The aim of this chapter has been twofold: first, to present some calculations that show the effect on the stratosphere, and on stratospheric ozone in particular, of the changing levels of carbon dioxide, methane, nitrous oxide and the CFCs; and, secondly, to place these calculations in context by discussing their limitations and uncertainties.

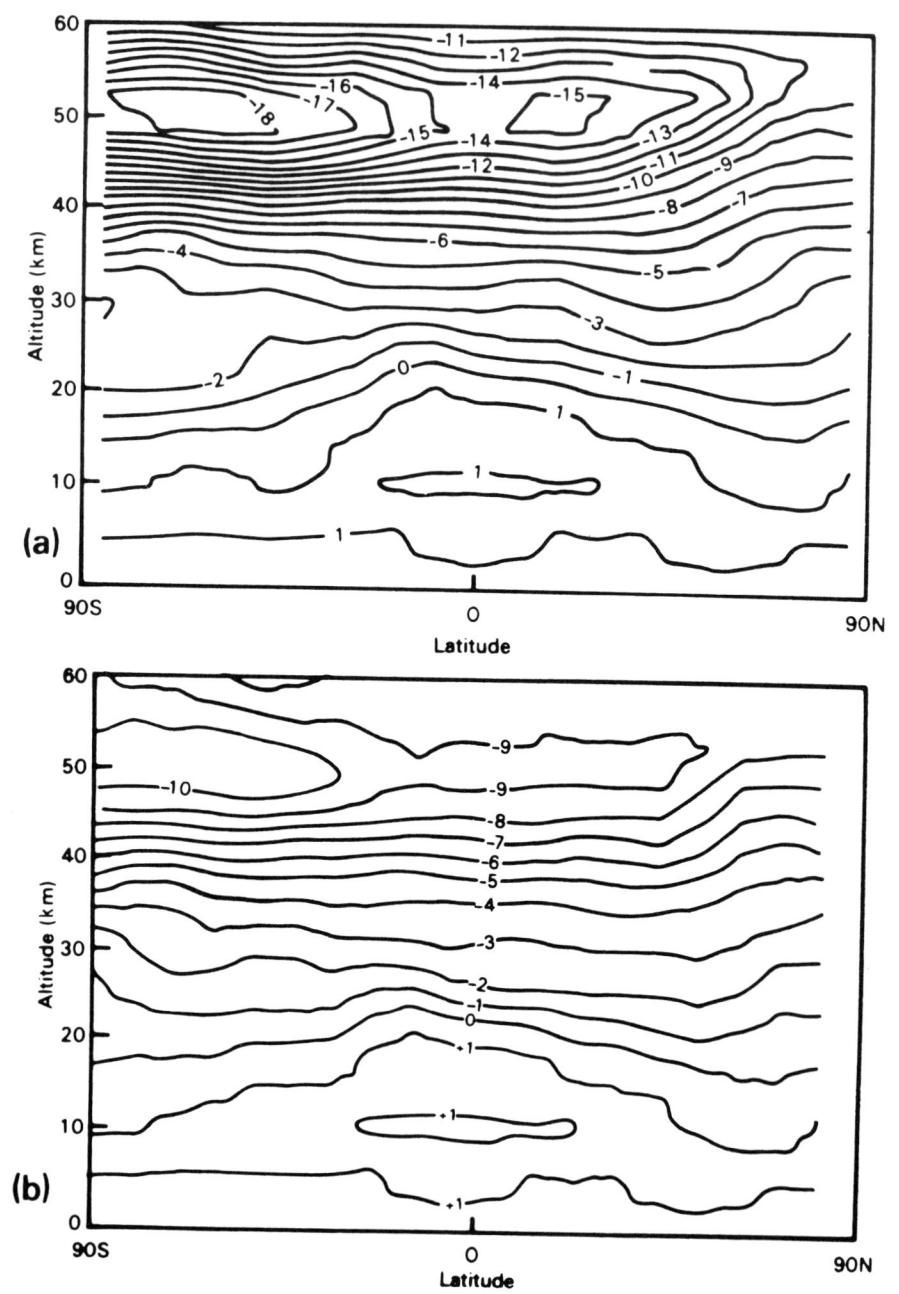

Figure 7 Temperature change (K) between 1985 and 2040 calculated with the Cambridge University model: (a) run CU4, (b) run CU7. (Reproduced from SORG[15] by permission of the Controller of Her Majesty's Stationery Office.)

Calculations assuming a continued growth in CFC emissions at 3% per annum predict a large build-up of active chlorine in the stratosphere to about 14 ppbv by 2040, with a global ozone depletion of nearly 7% compared with 1985. Even assuming emission rates that follow the Montreal Protocol, active chlorine levels will approximately double by 2040. This leads to a small global ozone depletion but with a significant percentage loss in the upper stratosphere. Our knowledge of the mechanism of the Antarctic ozone depletion (the 'ozone hole') indicates that, under the terms of the Protocol, depletion will continue. To stabilize stratospheric chlorine levels calls for emission reductions of about 85%. Larger reductions will be required to return to the Antarctic ozone levels of the early 1970s.

These calculations have many uncertainties associated with the formulation of the model and the input data (emission scenarios) used. The uncertainties due to model formulation include the uncertainty due to inadequately modelled chemical–dynamical–radiative feedbacks; the limitation in describing a three-dimensional spatial problem, with variability on all scales, by a coarse-resolution grid in one or two dimensions; and, possibly, the omission of important, but unknown or poorly understood, processes. A particularly important example of the latter is the omission of heterogeneous chemistry, now known to be extremely important in Antarctic ozone depletion, in these long-term integrations. It seems likely that, because these processes are not included, the ozone depletions presented here represent an underestimate.

Inadequate knowledge of emission data also severely limits the confidence in these calculations, although it does not alter the general conclusions concerning emission of CFCs into the atmosphere. How ozone will change globally during the next 100 years will depend critically on the combined, and sometimes compensating, effects of CFC (and halon) emission changes and the emissions of methane, nitrous oxide and carbon dioxide.

In all these predictions the role of the lower stratosphere is particularly important. The lower stratosphere is a difficult region to model. Improved understanding here should be a priority for further measurement and modelling activities.

ACKNOWLEDGEMENTS

This work was supported by the UK Department of the Environment.

REFERENCES

1. P.J. Crutzen, Ozone production rates in an oxygen–hydrogen–nitrogen oxide atmosphere. *J. Geophys. Res.,* **76**, 7311–7327 (1971).
2. H.S. Johnston, Reduction of stratospheric ozone by nitrogen oxide catalysts from supersonic transport exhaust. *Science,* **173**, 517–522 (1971).

3. R.S. Stolarski and R.J. Cicerone, Stratospheric chlorine: a possible sink for ozone. *Can. J. Chem.*, **52**, 1610–1615 (1974).
4. M.J. Molina and F.S. Rowland, Stratospheric sink for chlorofluorocarbons: chlorine atom catalyzed destruction of ozone. *Nature*, **249**, 810–814 (1974).
5. J.C. Farman, G. Gardiner and J.D. Shanklin, Large losses of total ozone in Antarctica reveal seasonal ClO_x/NO_x interaction. *Nature*, **315**, 207–210 (1985).
6. J.G. Anderson, W.H. Brune, S.A. Lloyd *et al.*, Kinetics of O_3 destruction by ClO and BrO within the antarctic vortex: an analysis based on *in situ* ER-2 data. *J. Geophys. Res.*, (1989), in press.
7. J.S. Barrett, P.W. Solomon, R.L. de Zafra *et al.*, Formation of the Antarctic ozone hole by the ClO dimer mechanism. *Nature*, **336**, 455–458 (1988).
8. World Meteorological Organisation, *Atmospheric Ozone 1985*. Global ozone research and monitoring project, report Number 16, Geneva, Switzerland (1986).
9. W.L. Grose, J.E. Nealy, R.E. Turner *et al.*, Modelling the transport of chemically active constituents in the stratosphere. In G. Visconti and R. Garcia, eds, *Transport Processes in the Middle Atmosphere*, NATO ASI Series, Series C, Mathematical and Physical Sciences, Vol. 213, pp. 229–250 (1987).
10. W.H. Brune, E.M. Weinstock and J.G. Anderson, Mid-latitude ClO below 22 km altitude: measurements with a new aircraft-borne instrument. *Geophys. Res. Lett.*, **15**, 144–177 (1988).
11. R.S. Eckman, J.D. Haigh and J.A. Pyle, An important uncertainty in coupled chlorine–carbon dioxide studies of atmospheric ozone modification. *Nature*, **32**, 616–619 (1987).
12. R.S. Harwood and J.A. Pyle, A two-dimensional mean circulation model for the atmosphere below 80 km. *Q. J. R. Met. Soc.*, **101**, 723–748 (1975).
13. J.A. Pyle, A calculation of the possible depletion of ozone by chlorofluorocarbons using a two-dimensional model. *Pure Appl. Geophys.*, **118**, 355–377 (1980).
14. J.D. Haigh and J.A. Pyle, Ozone perturbation experiments in a two-dimensional circulation model. *Q. J. R. Met. Soc.*, **101**, 723–748 (1982).
15. SORG, Stratospheric Ozone Review Group, UK, *Stratospheric Ozone, 1988*, HMSO, London (1988).

5
Present State of Knowledge of the Ozone Layer

ROBERT T. WATSON
Division of Earth Sciences and Applications, National Aeronautics and Space Administration, USA

ABSTRACT

There is now compelling observational evidence that the chemical composition of the atmosphere is changing at a rapid rate on a global scale. The atmospheric concentrations of carbon dioxide, methane, nitrous oxide, halons, and several chlorofluorocarbons are currently increasing at rates ranging from 0.2 to more than 5.0% per year. These changes in atmospheric composition in part reflect the metabolism of the biosphere and in part are due to national and international energy, agricultural and other industrial policies. Effecting a change in policy will require a nationally and internationally coordinated program of interdisciplinary research.

Information acquired by remote sensing provides a unique resource for verifying model experiments on the magnitude of, and rate of change in, the vertical distribution and total column content of ozone with latitude and season. Satellite data have been used to demonstrate that ozone in the spring over Antarctica has decreased significantly since the mid-1970s and that the decreases were not just confined to an area above Antarctica but extended from the South Pole to about 50°S. Both chemical and dynamical explanations have been advanced to explain the observations.

Ozone is predicted to decrease in the middle to upper stratosphere owing primarily to the increasing concentrations of chlorofluorocarbons, and to

---

Ozone Depletion: Health and Environmental Consequences
Edited by R. Russell Jones and T. Wigley
© 1989 John Wiley & Sons Ltd

increase in the troposphere due primarily to the increasing concentrations of methane. The impact of changes in ozone, and hence ultraviolet radiation reaching the Earth's surface, will be seen on human health, the productivity of aquatic and terrestrial ecosystems, and climate.

Long-term, continuous, calibrated data sets are needed for improvements in our documentation and understanding of global-scale changes in the Earth's environment.

ATMOSPHERIC OZONE

It is evident that the Earth is a planet characterized by change. We have entered an era in which the human race has achieved the ability to alter its environment on a global scale. The ozone and global greenhouse warming issues that are presently at the centre of attention of many scientists and policy-makers are just two of the interrelated environmental issues we face today.

To gain a scientific understanding of how human activities will affect the Earth's environment requires a new approach to Earth sciences. We need to obtain a scientific understanding of the entire Earth system on a global scale by describing how its component parts and their interactions have evolved, how they function, and how they may be expected to continue to change on all time scales. In particular, the immediate challenge is to develop the capability to predict the changes that will occur in the next decade and century, both naturally and in response to human activity. This will require a nationally and internationally coordinated program of interdisciplinary research to investigate long-term (10–100 years), coupled physical, chemical and biological changes in the Earth's environment, recognizing that land, atmospheric, oceanic and biospheric processes are strongly coupled on a variety of temporal and spatial scales.

This chapter briefly describes the current scientific understanding of the processes that control the abundance and distribution of atmospheric ozone, and its susceptibility to human-induced changes. Although the discussion focuses on the ozone issue, it is now widely recognized that the ozone depletion and greenhouse warming issues are strongly coupled because changes in ozone are predicted to modify the Earth's climate, and because the same gases that are predicted to modify ozone are also predicted to produce a climate warming.

BACKGROUND

For several decades, scientists have sought to understand the complex interplay among the chemical, radiative and dynamical processes that govern the structure of the Earth's atmosphere. During the last decade or so there has been particular interest in studying the processes that control atmospheric ozone, since it has been predicted that human activities might cause harmful effects to the environment by modifying the total column amount and vertical distribution of atmospheric ozone. Ozone is the only gas in the atmosphere that prevents harmful solar ultraviolet radiation from reaching the surface of the Earth. Unlike some other more localized environmental issues such as acid deposition, ozone layer modification (like global greenhouse warming) is a global phenomenon which affects the well-being of every country in the world. Changes in the total column amount of atmospheric ozone would modify the amount of biologically harmful ultraviolet radiation penetrating to the Earth's surface with potentially adverse effects on human health (melanoma and non-melanoma skin cancer, eye damage and suppression of the immune response system) and on the productivity of aquatic and terrestrial ecosystems. Changes in the vertical distribution of atmospheric ozone, along with changes in the atmospheric concentrations of other infrared-active (greenhouse) gases, could contribute to a change in climate on a regional and global scale by modifying the atmospheric temperature structure, which, in turn, could lead to changes in atmospheric circulation and precipitation patterns. The so-called greenhouse gases are gases that absorb infrared radiation emitted by the Earth's surface, thus reducing the amount of energy emitted to space, resulting in a warming of the Earth's lower atmosphere and surface (see Chapter 7 for further details).

The ozone issue and the greenhouse warming issue are strongly coupled because ozone itself is a greenhouse gas, and because the same gases that are predicted to modify ozone are also predicted to produce a climate warming. These gases include carbon dioxide (CO_2), methane (CH_4), nitrous oxide (N_2O), several chlorofluorocarbons (CFCs), including CFCs 11 ($CFCl_3$), 12 (CF_2Cl_2) and 113 ($C_2F_3Cl_3$), and halons 1211 (CF_2ClBr) and 1301 (CF_3Br). Methane, nitrous oxide, the CFCs and the halons, respectively, are precursors to the hydrogen, nitrogen, chlorine and bromine oxides that can catalyse the destruction of ozone in the stratosphere by a series of chemical reactions. Concentrations of these gases in the parts per billion range, control the abundance of ozone whose concentration is in the parts per million range; one molecule of a chlorofluorocarbon destroys thousands of molecules of ozone. Carbon monoxide (CO), which is not itself a greenhouse gas, and carbon dioxide can affect ozone indirectly. Carbon monoxide controls the concentration of the hydroxyl radical in the troposphere, which controls the atmospheric concentrations of some of the gases that can affect stratospheric chemistry. Carbon dioxide plays a key role in controlling the temperature structure of the stratosphere, which itself is important in controlling the rates at which the hydrogen, nitrogen, chlorine and bromine oxides destroy ozone.

There is now compelling observational evidence that the chemical composition of the atmosphere is changing at a rapid rate on a global scale. The atmospheric concentrations of carbon dioxide, methane, nitrous oxide and CFCs 11 and 12 are currently increasing at rates ranging from 0.2 to more than 5.0% per year. The concentrations of other gases, including CFC 113 and halons 1211 and 1301, important in the ozone and global warming issues, are also increasing, some at an even faster rate. These changes in atmospheric composition reflect, in part, the metabolism of the biosphere and, in part, a broad range of human activities, including agricultural and energy production practices. It should be noted that the only known source of the CFCs and halons is industrial production. They are used for a variety of purposes, including aerosol propellants, refrigerants, foam-blowing agents, solvents and fire retardants. At present, one of our greatest difficulties in accurately predicting future changes in ozone or global climate is our inability to predict the future evolution of the atmospheric concentrations of these gases. We need to understand the role of the biosphere in regulating the emissions of gases such as methane, carbon dioxide, nitrous oxide and methyl chloride (CH_3Cl) to the atmosphere, and we need to know the most probable future industrial release rates of gases such as the CFCs, halons, nitrous oxide, carbon monoxide and carbon dioxide, which depend on economic, social and political factors.

One important aspect of the ozone and global warming issues is that the atmospheric lifetimes of gases such as nitrous oxide, CFC 11 and CFC 12 are known to be very long. Consequently, if there is a change in atmospheric ozone or climate caused by increasing atmospheric concentrations of these gases, the full recovery of the system will take several tens to hundreds of years after the emission of these gases into the atmosphere is terminated.

MODEL PREDICTIONS

Numerical models are used as a tool to predict to what extent human activities will modify atmospheric ozone and climate. One-dimensional models are used to predict changes in the column content of ozone and the vertical distribution of ozone and temperature, but cannot predict variations in ozone or temperature modification with latitude, longitude or season. Major progress has been made over the past few years to develop two-dimensional models that can predict the variation of ozone and temperature change as a function of season and latitude. Three-dimensional models, which include longitudinal variations, are being developed to study the coupling between the chemical, radiative and dynamical processes that control the distribution of ozone and temperature, but these models are not yet ready to perform perturbation calculations.

Because it is now well recognized that the chemical effects of these gases on atmospheric ozone are strongly coupled and should not be considered in isolation, the most realistic calculations of ozone change take into account the impact of simultaneous changes in the atmospheric concentrations of carbon dioxide, methane, nitrous oxide, the CFCs and, possibly, other gases such as carbon monoxide, oxides of nitrogen (NO_x) and bromine-containing substances. The effects of these trace gases on ozone are not simply additive. Increased atmospheric concentrations of CFCs and nitrous oxide are predicted to decrease the column content of ozone, whereas increased atmospheric concentrations of carbon dioxide and methane are predicted to increase the column content of ozone. Therefore, the effects of increasing concentrations of CFCs and nitrous oxide are, to some degree, offset by increasing concentrations of carbon dioxide and methane. This is in contrast to the global warming issue, where increased atmospheric concentrations of the same trace gases are all predicted to increase the temperature of the atmosphere in an approximately cumulative manner.

One-dimensional model calculations have been performed to predict how ozone would change with time, assuming that the atmospheric concentrations of carbon dioxide, methane and nitrous oxide continue to increase at their current rates of 0.5, 1.0 and 0.2% per year, respectively, for the next 100 years, in conjunction with different assumptions for the annual growth rates in the emission of chlorine- and bromine-containing chemicals. Continued growth of CFCs and halons at 3% per year, which, in the absence of a ratified Montreal Protocol, is consistent with economic projections, is predicted to yield a globally averaged overhead column ozone depletion of about 6% by the year 2040 and more thereafter (see Chapter 4). This change is greater than natural variability and hence significant. In contrast, a true global freeze of the sum of world-wide emissions of all chlorine- and bromine-containing chemicals at or below projected 1990 levels, which, depending on the number of signatures and the growth rate in the non-signatory countries, may be consistent with the Montreal Protocol, is calculated to result in global column ozone depletions of less than 1% by the year 2015 and less thereafter. The results of these calculations demonstrate the strong chemical coupling that exists between these gases, and the time scale on which ozone changes are predicted to occur. In essence, when the growth rates of the CFCs are less than the growth rates of methane and carbon dioxide, only small column ozone changes are predicted because the CFC effects on ozone are temporarily masked. However, when the growth rates of the CFCs exceed those of methane and carbon dioxide, these gases can no longer buffer the impact of the CFCs and large column ozone depletions are predicted.

It should be noted that even when the predicted column ozone changes are small, as in the case of a true global freeze, and hence little change is expected in the amount of ultraviolet radiation reaching the Earth's surface, major changes in the vertical distribution of ozone are still predicted with potential consequences for climate. Ozone is predicted to decrease in the middle to upper stratosphere (approximately 25% near 40 km altitude) owing primarily to the increasing concentrations of CFCs, and to increase in the troposphere (up to 10% near the Earth's surface) owing primarily to the increasing concentrations of methane. The large predicted reductions in ozone near 40 km will lead to a local cooling of about 5°C. The consequences of this cooling for climate at the Earth's surface are currently unclear.

Two-dimensional models, which do not incorporate a dynamical feedback with ozone change, predict a significant variation in the ozone column decrease with latitude with the greatest depletions occurring at high latitudes. Depending on the exact trace gas scenarios used to predict ozone change, the pole-to-equator ratio of ozone depletion can range from a factor of 2 to 10. Seasonal effects are predicted but are somewhat less pronounced than the latitudinal effects. In general, two-dimensional models predict somewhat greater amounts of ozone depletion for the same trace gas scenarios than do one-dimensional models. For example, one two-dimensional model calculation for a true global freeze predicts a global average ozone depletion of about 2% during the next 50–70 years.

OZONE OBSERVATIONS

Global column ozone trends

A crucial question is to assess the extent of changes in global ozone that have already taken place, and to compare the changes to what has been predicted by theory. The search for global ozone trends involves looking for small secular changes amidst large natural variations that occur on many time scales. Observations of the total column content and the vertical distribution of ozone have been made for several decades using networks of different measurement techniques. Unfortunately, all of these observational techniques have certain limitations, which tend to restrict confidence in the results. These limitations arise from factors such as the lack of continuity of reliable calibration and the uneven geographical distribution of stations. Statistical analyses of the data are required to identify small trends among high natural variability, using data from relatively few stations. This chapter focuses on the long-term trends in column ozone.

NASA and the rest of the scientific community have long believed that it is imperative that all ground-based and satellite ozone data should be reanalysed and interpreted as soon as possible. It is crucial to evaluate whether the data have been analysed correctly and, if so, whether the decreases reported recently in total column ozone since the late 1970s are caused by natural events such as a decrease in solar radiation (from solar maximum to solar minimum), the 1982 eruption of El Chichon, or the 1982 El Niño event, or whether they are due to human activities such as the use of chlorofluorocarbons.

Therefore, during the fall of 1986, NASA, in conjunction with the Federal Aviation Administration (FAA), the National Oceanic and Atmospheric Administration (NOAA), the World Meteorological Organization (WMO) and the United Nations Environment Programme (UNEP), launched a major review of all ozone data. The scientific assessment report, *Report of the International Ozone Trend Panel: 1988* (WMO Report No. 18, 1989) covers:

(1) Calibration procedures and instrument performance
(2) Information content of algorithms

(3) Trends in total column ozone
(4) Trends in the vertical distribution of ozone
(5) Trends in stratospheric temperatures
(6) Comparison of observed and theoretically predicted trends of ozone
(7) Trends in source gases
(8) Trends in minor constituents
(9) Trends in aerosol abundances and distributions
(10) Observations and theories related to the Antarctic ozone hole
(11) Statistical procedures used to analyse trend data.

Total column ozone measurements have been made for more than 30 years at many locations around the world using ground-based Dobson spectrophotometers. Satellite measurements suitable for detecting long-term changes in ozone began only in the late 1970s. Hence, at present, long-term changes and trends over decades can be assessed only using ground-based data. The International Ozone Trend Panel found that all of the ground-based and satellite data had to be critically reanalysed before being used for the accurate determination of trends. The Dobson data were reanalysed taking more fully into account instrument calibration changes. Furthermore, data from certain stations were deemed unusable because of questionable quality. The accuracy of the re-evaluated data from a good Dobson station is estimated to be better than 0.7% per decade.

The quantity and quality of ground-based observations in the tropics and sub-tropics are such that the determination of trends is far less precise there than for northern mid-latitudes. In addition, reliable data are sparse in the southern hemisphere outside Antarctica. Thus, the Dobson data are not adequate to determine total column ozone changes in the tropics, sub-tropics or Southern Hemisphere. The satellite data, while providing global coverage, are affected by long-term drifts in instrument calibration. The ground-based stations furnish these needed calibration standards for assuring stable performance in the satellite assessment of contemporary global changes in total column ozone.

Total column ozone abundances at individual ground-based stations are known to be affected by at least three natural cyclic geophysical changes, namely the annual cycle of the seasons, the quasi-biennial oscillation (QBO) of the stratospheric winds, with a repeating cycle of about 26 months, and the 11-year solar sunspot cycle. There might also be natural effects from other irregular transient phenomena, such as the El Niño/Southern Oscillation (ENSO), or sporadic events such as volcanic eruptions. In addition, strong theoretical reasons exist for believing that the injection of the oxides of nitrogen from the atmospheric nuclear bomb tests of the late 1950s and early 1960s caused a reduction in total column ozone at that time. A statistical analysis of the ground-based ozone data from the Dobson stations was made to assess whether, after allowance for the perturbations from seasonal cycles, the QBO, the solar sunspot cycle, and nuclear tests, a residual linear trend existed in the measurements after

1970. Any residual trend can then be compared with theory to test our understanding of the impact of the increased abundance of trace gases or other poorly understood natural phenomena such as the effects of the volcanic eruption of El Chichon.

The analysis of the ground-based ozone data can be carried out either through examination of the individual results on a station-by-station basis, or by averaging the data over broad latitudinal bands in the Northern Hemisphere, where sufficient high-quality data are available. Both procedures have been used to analyse the data for the period 1965–86. Table 1 shows the magnitude of the residual linear trends for individual months that correspond to ozone changes since 1969 over several latitude zones with the effects of the QBO and solar cycle effects removed. The data were also analysed using only those values measured before the volcanic eruption of El Chichon and the large ENSO event of 1982. The residual linear trends obtained from this analysis were not significantly different from those obtained using data through 1986. Based on this result it appears that the trends are not significantly affected by these natural events. The magnitudes of the QBO and solar cycle effects are shown at the bottom of Table 1. The standard errors shown in Table 1 should be combined with an estimated systematic error for each latitudinal belt, which would be undetected by the statistical regression procedure, of 0.3% per decade (0.5% for the period 1969–86). This would increase the magnitude of the standard errors shown in Table 1 by about 0.2% or less.

The correlation of ozone change with the QBO is known to be dependent on latitude and this conclusion is confirmed in Table 1. The ozone response to the solar cycle varies from negligible to about 2% peak to peak for different latitude belts, with the largest column ozone values occurring at the sunspot maximum (winter 1979–80). The remaining linear trends show that there has been a measurable decrease in the annual average total column ozone of 1.7–3.0% in all latitude bands from 30° to 64° in the Northern Hemisphere from 1969 to 1986 (Table 1). The decreases are largest during the winter months (2.3–6.2%), averaged for December through March, inclusive. These contrast to the smaller changes in column ozone during the summer months (+0.4 to −2.1%), averaged for June through August, inclusive. Although there are insufficient stations to determine an accurate trend in the latitudinal band average of total column ozone at latitudes higher than 64°N, the limited data suggest that there has been a decrease comparable in magnitude to that observed between 53 and 64°N.

It should be noted that there are observational balloon-sonde data suggesting that *tropospheric* ozone may have increased at mid-latitudes in the Northern Hemisphere by as much as 1% per year over the last 20 years. Because about one tenth of the total column ozone is in the troposphere, this would be equivalent to an increase in total column ozone of about 2%, possibly offsetting a decrease in stratospheric ozone. The reliability and hemispheric significance of these tropospheric data is currently a subject of debate.

Table 1 Coefficients of residual linear trends after multiple regression analysis of reanalysed Dobson total ozone measurements collected into band averages (total percentage changes for 1969–86)

Month	Latitude band		
	53–64°N	40–52°N	30–39°N
January	−8.3±2.2	−2.6±2.1	−2.2±1.5
February	−6.7±2.8	−5.0±2.2	−1.2±1.9
March	−4.0±1.4	−5.6±2.3	−3.5±1.9
April	−2.0±1.4	−2.5±1.7	−1.7±1.3
May	−2.1±1.2	−1.3±1.1	−1.7±0.9
June	+1.1±0.9	−1.8±1.0	−3.3±1.0
July	+0.0±1.1	−2.2±1.0	−1.3±1.0
August	+0.2±1.2	−2.4±1.0	−1.0±1.0
September	+0.2±1.1	−2.9±1.0	−1.0±0.9
October	−1.1±1.2	−1.5±1.5	−0.9±0.8
November	+1.5±1.8	−2.4±1.3	−0.1±0.8
December	−5.8±2.3	−5.5±1.7	−2.1±1.1
Annual average	−2.3±0.7	−3.0±0.8	−1.7±0.7
Winter average**	−6.2±1.5	−4.7±1.5	−2.3±1.3
Summer average***	+0.4±0.8	−2.1±0.7	−1.9±0.8
QBO*	−2.0±0.6	−1.3±0.6	+1.9±0.6
Solar*	+1.8±0.6	+0.8±0.7	+0.1±0.6

*Per cycle minimum to maximum.
**Winter months include December to March, inclusive.
***Summer months include June to August, inclusive.
All uncertainties given in this table and throughout this report represent one standard error. The total uncertainty in the linear trends for total column ozone (1969–86) can be calculated by combining the standard errors given in this table with a systematic error of 0.5% by taking the square root of the sum of their squares. The uncertainties shown in this table increase by 0.2% or less.

Theoretical calculations of the expected changes in the total column ozone over the last 30 years were performed especially for this study. These calculations were carried out using two-dimensional models, which predict ozone changes with season and latitude. These calculations took into account variations in the solar output, the input of the oxides of nitrogen due to the nuclear bomb tests around 1960, and the increase in trace gas abundances, but they could not take into account the impact of the QBO, the ENSO, or the 1982 volcanic eruption of El Chichon, since there is inadequate understanding of how to model the various possible associated effects. These calculations predict that the total column ozone should have varied over the solar cycle by at least 0.7% and up to 2% and should have decreased between 30 and 60°N by about 0.5–1.0% in summer and 0.8–2.0% in winter from 1969 to 1986, in response to the increased abundances of trace gases. One model predicts anomalous increases to summer total column ozone that are inconsistent with other model results. The calculated changes in total column ozone caused by variations in solar output are broadly consistent with those observed. However, while the predicted decreases in total column ozone caused by the increased abundances of trace gases are consistent with the observed decreases in summer, the mean values of the predicted decreases are less than the mean values of the observed decreases in winter, especially at high latitudes. The residual linear trends in total column ozone might be caused wholly, or in part, by the increased abundance of trace gases, primarily CFCs.

Antarctic ozone changes

Important new observational evidence on ozone changes has recently been obtained. Data from a single Dobson instrument at Halley Bay (76°S, 27°W) has indicated a considerable decrease (greater than 50%) in the total column content of ozone above the Antarctic during the spring period (late August to early November) since 1957, with most of the decrease apparently occurring since the mid-1970s. Although it is clear from both the ground-based and satellite data that the abundance of ozone dropped rapidly after the late 1970s, given the high level of interannual variability in Antarctic ozone, it is difficult to draw strong quantitative conclusions about the rate and magnitude of these ozone decreases. Satellite measurements using both the Nimbus 7 Total Ozone Monitoring System (TOMS) and Solar Backscatter Ultraviolet (SBUV) instruments have verified this trend over the Antarctic since 1979 and have demonstrated the spatial and temporal variations in this feature. It can be seen from the satellite data that the ozone decreases are not just confined to an area above Antarctica but extend from the South Pole to about 50°S. Ozone changes of a comparable magnitude are not observed over the Arctic. There are several key features of the Antarctic ozone hole including the following:

(1) The development of the minimum occurs mainly during late August and the month of September.
(2) The rate of change of ozone within a given Antarctic spring appears to be increasing.
(3) The depth of the ozone hole exhibits significant year-to-year variability associated with the phase of the QBO.
(4) The region of low ozone lasted until late November/early December in 1987.
(5) In 1987, ozone was lower at all latitudes south of 60°S than in any previous year since measurements began. The zonal-mean total ozone amounts at 60, 70 and 80°S for September and October of 1987 were between 75 and 150 Dobson units lower than the corresponding months of 1979.
(6) The ozone in 1988 was significantly less depleted than in 1987.
(7) The depletion extends from the South Pole to about 50°S.

(8) While the ozone depletion is largest in the Antarctic spring, ozone has decreased at all latitudes south of 50°S throughout the year.
(9) In some instances, total column ozone changed dramatically (25–50 Dobson units) over areas as large as several million square kilometres within a matter of a day or two, with such changes lasting for up to a few days.
(10) Balloon ozonesonde and SAGE satellite data indicate that the ozone decrease is confined to a region between about 12 and 24 km.
(11) In October of 1987, ozone concentrations fell to less than 5% of their August values between about 15 and 20 km.

It is not yet evident whether the behaviour of ozone above the Antarctic is an early warning of future changes in global ozone or whether it will always be confined to the Antarctic because of the special geophysical conditions that exist there. Until the processes responsible for the decrease in spring Antarctic ozone are more fully understood, it will not be possible to state with any certainty whether it is a precursor of a trend in the Southern Hemisphere.

Three basic theories have been proposed to explain the observed decrease in spring Antarctic ozone that has been occurring since the late 1970s:

(1) The hole is caused by the human activity of increasing the atmospheric loading of chlorinated and brominated chemicals (CFCs and halons). These compounds could then efficiently destroy stratospheric ozone in the Antarctic environment because of the special geophysical conditions that exist in this region of the atmosphere.
(2) There have been changes in the circulation of the atmosphere, which now transports ozone-poor air into Antarctica.
(3) Periodically enhanced abundances of oxides of nitrogen produced by solar activity can cyclically destroy ozone.

A major ground-based field measurement campaign took place at McMurdo between August and November of 1986 to study the ozone layer above the Antarctic. This campaign was co-sponsored and coordinated by NASA, NOAA, the US National Science Foundation (NSF) and the Chemical Manufacturers Association (CMA). The campaign was exceedingly successful and the data from all four groups of experimentalists have now been published. An extensive set of measurements of the chemical composition was obtained, including hydrochloric acid, chlorine nitrate, chlorine monoxide radical, chlorine dioxide, nitric acid, nitrogen dioxide, nitric oxide, nitrous oxide, hydrofluoric acid, ozone and other chemical species. From the data it is clear that the chemical composition of the lower stratosphere over Antarctica is significantly perturbed relative to that expected from theoretical considerations using standard homogeneous chemistry. The abundances of odd nitrogen species are low, while the partitioning of odd chlorine species is significantly different from that observed at mid-latitudes.

However, from that data alone it was premature to conclude that the cause of the loss of ozone over Antarctica is due to chlorofluorocarbons.

Consequently, although the 1986 campaign was a major success, producing excellent data on the chemical composition of the spring Antarctic atmosphere, it was clear that additional scientific data were required for further refinement of our understanding of the processes controlling ozone in and around Antarctica. Therefore, NASA, NOAA, NSF and the CMA mounted a major campaign in 1987 using ground-based, aircraft and satellite instrumentation. It consisted of two components: the 1987 ground-based campaign, coordinated by NSF at McMurdo, which was somewhat expanded in scope compared to the 1986 NOZE I campaign; and a 1987 aircraft campaign, organized and managed by NASA on behalf of the four agencies. The latter used the NASA ER-2 and DC-8 aircraft as platforms for instruments operated by scientists from NASA Centers, the NOAA Aeronomy Laboratory, the National Center for Atmospheric Research and universities, together with meteorological satellite data.

These aircraft were equipped with state-of-the-art instrumentation to determine the chemical composition and physical state of the atmosphere. The scientific payloads for both the ER-2 and the DC-8 were selected to critically test the proposed CFC, solar and dynamical theories. In addition, the scientific payloads were designed to obtain a broad base of scientific data in case all currently proposed mechanisms were incorrect. The campaign, which involved over 150 scientists, engineers and pilots, was based in Punta Arenas, Chile, and ran from mid-August until the beginning of October.

The processes controlling the abundance and distribution of ozone in Antarctica are complex and intertwined. However, given the successful nature of the 1987 campaign, the scientific community is now in a position to start to appreciate more fully the exquisite balance between the meteorological motions and the photochemistry.

The weight of evidence strongly indicates that man-made chlorine species are primarily responsible for the observed decrease in ozone within the polar vortex. However, it must be recognized that the unique meteorology during winter and spring over Antarctica sets up the special conditions of an isolated air mass (polar vortex) with cold temperatures required for the observed perturbed chemical composition.

It is quite evident that the chemical composition of the Antarctic stratosphere is highly perturbed compared with elsewhere on the planet. The aircraft findings are consistent with the observations made in 1986 from McMurdo, as well as those from the 1987 McMurdo expedition. The distribution of chlorine species is significantly different from that observed at mid-latitudes, as is the abundance and distribution of nitrogen species. The amount of total water within some regions of the vortex is significantly lower than anticipated.

From late August until the end of September, the abundance of the chlorine monoxide (ClO) radical within the chemically perturbed polar region was elevated by a factor of more than 100 relative to that measured at mid-latitudes at the highest altitude at which the ER-2 was flown (about 18.5 km). The abundance of ClO was observed to decrease towards lower altitudes. At the highest flight levels, the abundance of ClO at local solar noon reached, and on occasion exceeded, 1 ppbv. In addition to the decrease in ClO abundance at lower altitude, the abundance of ClO was also observed to decrease dramatically outside the chemically perturbed region.

Chlorine dioxide (OClO), which is most likely formed in a reaction sequence involving the ClO radical, was observed both day and night at highly elevated concentrations compared to those at mid-latitude. The column content of hydrochloric acid, which is one of the major chlorine reservoirs at mid-latitudes, is very low within the chemically perturbed region, indicating that most of this inactive chlorine reservoir species has been converted to a more active form of chlorine below 25 km. The abundance of chlorine nitrate ($ClONO_2$) was observed to be elevated near the edge of the polar vortex, consistent with the observed abundances of ClO and nitrogen dioxide.

Thus, there is no longer debate as to whether ClO exists within the chemically perturbed region at ER-2 flight altitudes (15–19 km) at abundances sufficient to destroy ozone, if our current understanding of the chlorine-ozone catalytic cycle is correct. The rate of decrease in ozone during the month of September at most of the altitudes at which the ER-2 was operated during this campaign is broadly consistent with simultaneously observed concentrations of ClO. However, present understanding of key chemical reaction rates and photodissociation products within the catalytic process is still incomplete.

There is another line of observational evidence consistent with ozone destruction by chlorine catalysis. Between the end of August, when there was no strong correlation between ClO and ozone, and the middle of September, as the ozone concentration was dropping at ER-2 altitudes, a strong anticorrelation developed between ClO and ozone. This anticorrelation is consistent with the picture of ozone destruction by ClO.

The bromine monoxide (BrO) radical, which has been implicated in some of the proposed CFC-halon theories, has been observed at concentrations of a few pptv within the chemically perturbed region of the vortex at the flight altitudes. The low measured abundance of bromine monoxide means that, although bromine is involved in the chemical destruction of ozone, it is not involved in the primary chemical mechanism.

The ER-2 observations of the abundance of odd nitrogen (NO_y), which is the sum of all nitrogen-containing reservoir and radical species, show, like total water, very low values within the chemically perturbed region of the vortex, indicating that the atmosphere has been denitrified, as well as dehydrated. Abundances of NO_y of 8–12 ppbv were observed outside the chemically perturbed region, while abundances of 0.5–4 ppbv were observed inside the chemically perturbed region. A similar large change was observed for one of the nitrogen components, nitric oxide. In addition, some of the NO_y observations suggest that NO_y component species are incorporated into polar stratospheric cloud (PSC) particles and nitrate was observed in the particle phase on some of the filter samples and on some of the wire impactor samples taken in the chemically perturbed region of the vortex. The column measurements of nitric oxide, nitrogen dioxide and nitric acid made from the DC-8 exhibit a strong decrease in the abundance of these species towards the centre of the vortex. These low values of nitrogen species are contrary to all theories requiring elevated levels of nitrogen oxides, such as the proposed solar cycle theory.

Observational data that air within the chemically perturbed region of the vortex is dehydrated and that the NO_y abundances are very low are consistent with theories in which the chlorine reservoir species, $ClONO_2$ and HCl, react on the surfaces of polar stratospheric clouds to enhance the abundance of active chlorine species, i.e. ClO. The observations also support the picture that the abundance of NO_y is low because odd nitrogen can be removed from the atmosphere by being tied up in ice crystals, which can then gravitationally settle to much lower altitudes. Low abundances of NO_y are needed to prevent the rapid reconversion of ClO to $ClONO_2$. This picture is further supported by the observations of low column abundances of hydrogen chloride, by occasional observations of high levels of nitrate found in the ice particles, and by the visual and lidar observations of high cirrus and polar stratospheric clouds.

The data offer no support for sustained large-scale upwelling. In the restricted region covered by the ER-2, from 54 to 72°S and from altitudes of 12.5–18.5 km, measurements of CFC 11 and nitrous oxide, which act as tracers of air motions, show no evidence of a general increase in abundances above about 14 km during the mission, although there were instances of structure and elevated values.

Sudden changes in ozone were observed on several occasions. One such event was observed on 5 September over the Antarctic Peninsula, illustrating the complexity of the ozone hole. The magnitude and rapidity of the decrease cannot be ascribed to a chemical cause. Air of low ozone content appears to have been transported into the region. The origin of that air is not completely known. It could be either tropospheric/lower stratospheric air naturally low in ozone from mid-latitudes, or air in which ozone had been chemically depleted from within the polar vortex. Although these events are clearly not the cause of the ozone hole, they are important insofar as they are associated with an upwelling of the air. Thus, the air cools rapidly leading to the formation of PSCs, which then perturb the chemical composition of the air, i.e. the production of ClO from HCl and $ClONO_2$. The ClO produced can then lead to a chemical destruction of ozone.

MODEL RELIABILITY

A key question still remains concerning the reliability of the models used to predict ozone change. Given that we cannot directly test the accuracy of a prediction of the future state of the atmosphere, including the distribution of atmospheric ozone, we must test the models by trying to simulate the present atmosphere, including the distribution of atmospheric ozone, or by trying to simulate the evolution of the atmosphere, in particular ozone, over the past few years. This is done by comparing model predictions with atmospheric observations.

We should note that nearly all the key chemical constituents that are predicted to be present in the atmosphere, and that are important in ozone photochemistry, have now been observed. In general, the models predict the distribution of the chemical constituents quite well. However, the measurements are not adequate for critically testing the reliability of the photochemical models. Close examination of the intercomparison of measurements and model simulations of the present atmosphere reveals several disturbing disagreements. One of the major disagreements appears to be that modelled ozone concentrations are typically 20 — 50% lower than measured ozone concentrations in the upper stratosphere, where it should be easiest to predict the concentration of ozone, and where chlorine is predicted to have its maximum effect. These types of disagreement limit our confidence in the predictive capability of these models.

In the end, however, our predictive capability will be tested by measuring the changes taking place in the atmosphere. This will require careful measurements of critical species to be carried out over long time periods, i.e. decades. NASA, NOAA and CMA recently co-sponsored a workshop to design a 'Network for the Detection of Stratospheric Change'. This network would be designed primarily to provide the earliest possible detection of changes in the chemical and physical structure of the stratosphere, and the means to understand them. It would also provide an invaluable data set of latitudinal and seasonal variations in stratospheric chemical composition to test the validity of multidimensional models. Implementation of such a network is a high priority.

CLIMATE

As stated earlier, the observed increases in the atmospheric concentrations of the CFCs, methane, carbon dioxide and nitrous oxide also have direct implications for the Earth's radiative balance through the so-called greenhouse effect. These gases absorb infrared radiation in a part of the spectrum which is otherwise transparent. Presently, and in the near future, changes in the concentrations of trace gases other than carbon dioxide are thought to be contributing to the greenhouse warming of the Earth's surface and lower atmosphere by an amount that is about equal to that due to changes in the concentration of carbon dioxide. The cumulative effect of the increase in all trace gases for the period 1850–1988 is a predicted equilibrium warming (i.e. ignoring the lag or damping effect due to the heat capacity of the oceans) in the range 0.7–2.2 K, about half of which should have occurred to date. Model calculations indicate that the greenhouse warming during the next 50 years should be about 1.2–2.3 K (see Chapter 8), substantially more than that which has occurred during the previous 130 years. Thus, problems of ozone change and climate change should be considered together. It is also apparent that what has been previously thought of as the carbon dioxide–climate problem should more properly be thought of as the trace gas–chemistry–climate problem.

The key issues are the magnitude and timing of the predicted global warming. The magnitude is not only sensitive to the rate of change of the atmospheric abundances of the trace gases, but also depends upon future changes in cloud cover and distribution. The timing of the warming is largely controlled by the equilibrium sensitivity of the climate system and the rate at which heat is mixed into the oceans (i.e. by the thermal inertia of the oceans; see Chapter 8).

SUMMARY

The composition of the atmosphere is now changing rapidly because of a variety of influences of both natural and human origin. These changes have implications for a variety of problems and, in particular, demonstrate the connections among the studies of global tropospheric and stratospheric chemistry, trace gases and climate. The Antarctic ozone depletion phenomenon vividly demonstrates that the environment does not always change slowly, linearly or predictably in response to a perturbation. Consequently, we must realize that a global-scale experiment on the atmosphere of planet Earth is being conducted by humankind without our fully understanding the consequences.

It is clear that significant progress has been made in our understanding of the physical and chemical processes that control the distribution of ozone and the temperature structure of the atmosphere. However, we must recognize that significant uncertainties in our knowledge remain, and that these can be resolved only by a vigorous programme of research.

Discussion Period 1

Q I expect you are familiar with the paper published in *Science* earlier this year which presented the results of ground level ultraviolet measurements in the USA from 1974 to 1985. The data showed a decrease in ground level ultraviolet of about 0.7% per year, which seems incompatible with a decrease in ozone even of the modest levels which you have suggested.

Joe Farman The measurements to which you refer are made at airfields and the authors of the *Science* article do not comment on any changes in air traffic. One would surmise that, if air traffic increased, the extra soot in the atmosphere could screen out the amount of sunlight at ground level, so the ground-based meters are not necessarily measuring changes in stratospheric ozone.

Q If, as predicted, troposphere ozone increases as stratospheric ozone declines, what will happen to the absorption of ultraviolet radiation? May not some penetrate and reach people who are skiing or living at high altitude?

Professor Isaksen No. The levels at which these changes are occurring is well above the level of human habitation. You must remember that penetration of ultraviolet radiation is determined by total column ozone; that is, the stratosphere and troposphere combined. Of course, tropospheric ozone may increase at a lower altitude as a result of air pollution, but that is a separate problem.

Q Can you give us some indication as to the stability of your models and their sensitivity to the kinetic data that go into them?

Dr Pyle Professor Isaksen made this point very well for CFCs. Changes in the kinetic data used in models have changed the predictions by a factor of 2 over about 10 years. For high-flying aircraft, changes in kinetic data actually changed the sign of ozone depletion at one stage. Since the early 1970s, however, there has been a tremendous amount of laboratory study. We ought now to be more immune to changes in the kinetic data of the kind that were prevalent in the past. A more important source of uncertainty is the precise chemical scheme that governs ozone depletion, as we have seen in relation to Antarctic ozone.

Professor Isaksen I think we have a fairly good knowledge of gas phase chemistry. The uncertainty relates to the heterogeneous reactions and whether the processes in the Antarctic are relevant elsewhere. This could alter our modelling predictions and would make the ozone depletion problem much worse.

Dr Pyle About a year ago, I published some modelling studies in *Nature* using new kinetic data with different temperature dependence for some of the hydrogen radical reactions. That changed the predictions so that the greatest changes no longer occurred at the highest latitudes.

Professor Isaksen I would just add that there are large uncertainties connected with the transport of ozone and the way that it is incorporated into the models.

Q Your models show very significant changes in the temperature within the stratosphere. Is this likely to affect meteorological conditions, and should this be included in the models?

Dr Pyle This is an important point. It is very difficult to predict exactly what will happen in a model of the kind described. The one we have been using is not a detailed meteorological model. However, according to groups like GFDL, even if you halve ozone you still find that the responses are mainly radiative; that is to say, the temperature changes but the meteorology, i.e. transport, doesn't change as much. Of course those experiments were very idealized, and there is still potential for meteorological changes and thus for changes in the transport of ozone.

Q What atmospheric data are being gathered to establish whether heterogeneous reactions are occurring on a global scale?

Dr Pyle There is research going on in the USA and there are people in the UK who are keen to become involved in this area. First you have to characterize the surface. In the Antarctic, ice crystals containing nitric acid and maybe hydrogen chloride are thought to be important, but we need to make measurements *in situ* to characterize them. Secondly, and this is important from a global point of view, are these reactions occurring on other surfaces, such as sulphate aerosol which we know is ubiquitous in the stratosphere? We do not have an answer to that yet. Laboratory studies will help but I suspect that such heterogeneous reactions will be even more difficult to characterize than ones on ice crystal surfaces. The research effort involved is enormous but it must be done.

There will be an expedition this winter to the Northern Hemisphere equivalent to that which visited Antarctica last year. It will be based in Norway and the main objective will be not just to look for evidence of ozone depletion at high latitude but to establish whether this type of reaction occurs over a large area outside the polar regions.

Professor Brian Thrush Ten years ago John Noxon reported a sharp drop in the nitrogen dioxide concentration around the North Pole in the spring. This became known as the Noxon cliff. Do you think there is early evidence of an ozone hole in the arctic?

Bob Watson Not exactly. I believe that the Noxon cliff is evidence that the chemistry at high northern latitudes is disturbed relative to mid-latitudes and equatorial latitudes. It is likely that nitrogen compounds get tied up, probably as N_2O_5, and possibly the N_2O_5 is converted on the ice crystals into nitric acid, which might then precipitate out of the atmosphere. One problem is that the ice crystals in the northern hemisphere probably don't grow large enough to precipitate out efficiently, i.e. they may not grow larger than $2\mu m$. Therefore the Noxon cliff is simply evidence for very low concentrations of nitrogen dioxide. This is evidence of unusual chemistry, but it does not provide definite evidence of an ozone hole.

Q You said that bromine contributes up to 25% of the ozone changes in Antarctica. What about the changes you observed over northern latitudes?

Bob Watson We don't understand exactly why there have been changes in the Northern Hemisphere from a quantitative point of view. The observed changes are typically 2 to 3 times greater than we would predict are due to CFCs. Theoretically one would expect the concentration of bromine to be very low relative to chlorine. On a per molecule basis bromine compounds are typically a factor of 10 more efficient than chlorine. However, the Earth's atmosphere contains only a few parts per trillion of bromine *versus* a few parts per billion of chlorine. So I would estimate that the bromine contribution was somewhere in the 2–10% range — say about 5%.

Q It was stated that El Chichon caused a perturbation in the ozone globally. What about the effects of future volcanic eruptions in a modified atmosphere? Is the atmosphere going to be more sensitive to volcanic effects now that this chlorine overburden exists? And is that not an area where future research should be targeted?

Bob Watson As you saw from the data, ozone seemed to reach a minimum just after the El Chichon eruption. This volcano ejected material directly into the stratosphere, whereas the material from Mount St Helens did not penetrate the stratosphere. Therefore, it is mainly volcanic eruptions that are explosive enough to eject material, gases and dust, directly into the stratosphere that are likely to affect ozone. Analysis of the data indicates that the reduction of ozone after the eruption of El Chichon was genuine. This raises the question of whether liquid droplets of sulphuric acid act as efficient surfaces for heterogeneous chemical reactions to occur. Talbert and Golden at SRI have studied this problem and they believe that sulphuric acid liquid droplets can allow heterogeneous processes to occur which would enhance the catalytic efficiency of chlorine by converting HCl to ClO. Thus we need to monitor not only the homogeneous chemical composition of the stratosphere, but the heterogeneous composition as well. In addition, we need more laboratory studies to quantify the mechanism by which chlorine gases and nitrogen gases can react on the surface of cold sulphuric acid droplets.

Q What are the principal biological consequences of localized ozone depletion, and has ultraviolet radiation been measured directly underneath ozone holes?

Joe Farman Detailed measurements have not yet been done underneath ozone holes. However, we can do the calculations, which are probably quite sufficient.

Bob Watson The National Science Foundation in the USA has provided funds to look specifically at the amount of ultraviolet radiation reaching the Earth's surface. Of course there is no historical record, so it is going to be difficult to compare it with previous years. Research is also needed to examine the impact of ultraviolet radiation on aquatic organisms in the oceanic regions of the Southern Hemisphere. One of the main problems in the USA, and I would also say within Europe, is that the amounts of money being spent on the impact of UVB radiation on ocosystems are wholly inadequate. This is a situation that we are going to live to regret in my opinion.

Professor Brian Thrush At the time that the ozone hole occurs, the sun is still very low in the sky, and therefore the path of ultraviolet light through the atmosphere is longer. Presumably that helps to mitigate the biological effects.

Bob Watson That is partly true but there are two other factors to take into consideration. One is the depth of the ozone hole and the other is its persistence. As Joe Farman showed in both 1985 and 1987, the ozone hole was not only deeper, it also lasted longer than in previous years. I think the reason for that is that ozone falls so low, there is less radiative heating of the atmosphere by ozone, so the vortex stays tighter and thus lasts longer. In 1987, the ozone hole did not really disappear until the first few days of December. By then it was almost midsummer in Antarctica, so the sun was quite high in the sky. In addition, the ozone hole moved away from the polar region to the oceanic regions outside of the icecap. This year (1988) the ozone hole wasn't as deep as in 1987. It was much shallower, but it was centred well away from the pole so that parts of the ozone hole were always over open ocean towards South America. One would have to question what the biological impact was, even this year, with a relatively shallow ozone hole.

Q CFCs contain fluorine. What is the impact of fluorine?

Bob Watson There is no concern from the point of view of ozone depletion. Fluorine atoms liberated from a CFC quickly react with any hydrogen-containing molecules in the atmosphere, such as molecular hydrogen, methane or hydrogen peroxide to form an H–F bond. The H–F bond is one of the strongest known chemical bonds and cannot be photolysed even with high doses of vacuum-ultraviolet light. Also, HF doesn't react with hydroxyl radicals, so once the HF molecule is formed it is extremely stable. Now, if fully fluorinated substitutes replace CFCs 11 and 12, although this will greatly benefit the chemical destruction of the ozone layer, the infrared properties of these chemicals are not so benign. They will have an almost infinite chemical lifetime and their greenhouse potential must be considered. I believe Keith Shine will talk about this later.

Q Could I ask about the very sharp edge of the ozone hole? It is very striking, either you have ozone or you have ClO. Is this essentially a temperature effect? As the ozone starts to deplete, it becomes colder, there is more condensation, and there is a positive feedback. Do you think this switch is primarily a temperature effect or a chemical effect, or both?

Bob Watson I think it is a combination. The meteorology sets up a contained vortex and the edge of the vortex seems to be contained by the polar edge of the jetstream. Within that polar vortex there are quite cold temperatures. The cold temperatures causes the polar stratospheric clouds to form; stratospheric clouds lead to the unusual chemistry. There is very little mixing across the vortex region.

Joe Farman I think there is a bit more to it than that. The size of the vortex is decreasing throughout the spring season. Parts are shredded off rather like taking skin off an onion. When an aircraft flies through a bit that has just been eroded, a sharp edge is suddenly created. If you could follow that air around it would gradually weaken in character I suspect.

Professor Brian Thrush There have been atomic bomb tests since the late 1940s. Was there any observable effect of those, or were there insufficient records? If there hadn't been a nuclear test bomb treaty in the early 1960s what would have happened?

Bob Watson Regarding the first question there were very few stations that were accurately measuring ozone before 1958. 1958 was the International Geophysical Year, and, starting that year, there was a proliferation of high-quality Dobson stations measuring ozone over the globe. Before this I do not know whether the few stations available would have produced data which could have distinguished the impact of the atomic bomb tests from natural variability. Also, I do not know what the total input of NO_x would have been from the earlier bomb series, or whether it would have been sufficient or comparable to the 1958–62 period. My guess is that there would have been insufficient activity from the earlier nuclear bomb tests to observe any effect.

The second point is what would have happened if there hadn't been a nuclear bomb test treaty. The injection of NO_x directly into the stratosphere is predicted to decrease ozone. This was observed in the early 1960s and would have continued or worsened. In general, we should be extremely careful about putting NO_x into the Earth's stratosphere. I know that certain aircraft companies are now reconsidering the viability of supersonic transport and are considering hypersonic transport flying as high as 30 km. Unless the engines have changed substantially since this idea was first proposed in the early 1970s we will again be faced with the issue of ozone depletion due to supersonic transportation. This needs to be looked at very very carefully before there is massive investment in supersonic transportation by the aerospace industry.

Q In order to assess the impact of increased UV radiation on the biosphere, better UV measurements are clearly going to be needed. Robertson–Berger meters record weighted data in sunburn units. What efforts are being made to obtain unweighted spectral data?

Bob Watson This is being done, but I think it is going to be very difficult to determine absolute trends in ultraviolet irradiance at different wavelengths. I say this for two reasons: the natural variability and the calibration of the instrument. To measure solar irradiation one needs direct absolute measurements of ultraviolet reaching the ground. This is much more complex than analysing the Dobson record, which measures the ratio of solar radiation at two wavelengths. In this situation most of the calibration problems disappear. The calibration problems of monitoring absolute ultraviolet at the ground are much harder to resolve. The second problem is interpretation, and the measurement of trends in the factors other than ozone which can affect penetration of ultraviolet light to ground level, e.g. clouds and local pollution.

Using Robertson–Berger meters, Scotto found no increase in UVB at sites in North America, but these meters were located at airports, and I am sure that local pollution and increase in traffic obscures any changes in UVB due to ozone changes. I would estimate that absolute trends won't be identified reliably until 20 or 30 years of data have been gathered.

Robin Russell Jones I cannot let either of you go without asking a crucial question. Assuming that the group 1 and 2 substances in the Montreal Protocol are phased out within a reasonable period of time, what dangers do you see to stratospheric ozone from the substitute chemicals and should these be controlled also?

Bob Watson I think it is absolutely essential that we look at the potential environmental impacts of all the proposed substitutes. I don't think there is any question about that. Policy-makers have to be pragmatic to a certain extent. It is difficult to envisage a society without refrigeration. The key questions for policy makers, therefore, are what is the key loading of chlorine that is acceptable, if any, and can we then calculate how much of a particular substitute can be safely released into the stratosphere? I would imagine that pragmatism will win out and that chemicals with an ozone depleting potential of less than 0.05 will be deemed acceptable, at least for the time being. In the short term that will reduce the threat of ozone destruction by a factor of 20. But it is also imperative that we understand why carbon tetrachloride is increasing, and that we monitor methylchloroform. While methylchloroform has an ozone depletion potential of only around 0.1 or 0.2, it is produced industrially in large quantities, and it makes no sense to regulate CFCs 11, 12 and 13 even to the 95% level, if methylchloroform and carbon tetrachloride are not controlled. All the other substitutes should be examined, both from the point of view of ozone depletion and greenhouse warming. Let's not forget that CFCs have two environmental impacts.

Joe Farman We haven't mentioned HCFC 22. The situation here is a bit more equivocal. It is supposed to have an ozone depleting potential of 0.05, but it does get awfully high in the atmosphere. Also, it does depend on the model you actually use to calculate the ozone depleting potential. These are not fixed values. The one lesson I have learned from Antarctica is that we certainly do not understand the chemistry of ozone well enough to predict the future with any confidence. There are still many uncertainties about the way in which we overturn the upper stratosphere and bring down the CFC decomposition products into Antarctica. There is no current model which handles that well, so we need to be very careful before declaring that a particular chlorine-bearing substance is safe, however low its ozone depleting potential might be using current models.

John Pyle Really just a 'rider' to the question Robin asked about the effects on the stratosphere. The alternative gases are going to react in the troposphere with hydroxyl. They are all organics, which are eventually going to end up as carbon monoxide, and carbon monoxide is the major sink for hydroxyl radicals. So if we allow these compounds to increase by 3% per year, how long before there is a tropospheric signal?

Bob Watson Is it not correct that the amount of carbon monoxide generated from these gases is small compared with other sources? Theoretically these gases will aggravate the acid rain problem, but again the amount is insubstantial compared with natural hydrogen chloride from sea salt and other man-made sources of sulphuric acid and nitric acid.

Part 3
Global Warming

6
Opening Remarks

JOHN MADDOX

Before introducing the first speaker I would like to say a few words about the subject of this afternoon's session: global warming. I wish to clarify my own position on this issue – not least because in the past I have sometimes been accused of being less than friendly towards environmentalists and the issues they espouse. However, the greenhouse effect is an issue that we ignore at our peril — certainly it is a field in which there are many uncertainties, which are unlikely to be resolved quickly. I would not be surprised if by the end of the century it will still be hard to make accurate predictions. Even so, it is exceedingly unlikely that the accumulation of carbon dioxide and other gases in the atmosphere can continue as it is at present without, at some point, there being climatic change, the end-point of which might be the melting of the ice caps. It is my opinion that this prospect is so serious that, whatever the uncertainties in the scientific predictions or modelling may be, doubts about how much carbon dioxide is recycled through the biosphere and so on, it is essential to begin soon to work out a framework in which it will be possible at some point in the future to control the way in which carbon dioxide and other gases are allowed to accumulate in the atmosphere. In many ways the Montreal Protocol is a valuable precedent for the kind of convention that will in due course have to deal with greenhouse gases. Obviously the problem of global warming has much greater implications economically. Even so, and despite the uncertainties that persist, it is by no means too soon to begin embarking now on the negotiation and the kind of agreements that will be necessary to regulate the greenhouse problem. Of course, this may prove quite difficult. It will not be simple to prevent countries by mutual agreement from releasing carbon dioxide and other materials into the atmosphere. Whereas the

---

Ozone Depletion: Health and Environmental Consequences
Edited by R. Russell Jones and T. Wigley
© 1989 John Wiley & Sons Ltd

CFCs that lead to the depletion of ozone are relatively unimportant economically, the control of carbon dioxide and other greenhouse gases will require much harder decisions to be taken.

Paradoxically, the negotiations should begin even before the scientific side of the problem has been fully resolved. The organizer of this Conference is anxious that I shouldn't use this occasion as an opportunity to promote the building of nuclear power stations. That is a very interesting question, but it is only part of the equation. Let me just say this: if there is an international agreement, based, for example, on national quotas of allowable carbon dioxide release per year, then every country will be free to respond to this challenge in its own way. We have learned enough about energy consumption in the past 15 years to know that there is no one recipe that holds for all time in any country, and indeed that countries differ enormously in the extent to which they rely on fuel efficiency, on nuclear power, or on renewable energy. Indeed it may even be possible to meet quota requirements by scrubbing carbon dioxide from the gases in power stations. The method used is not the most important thing. It seems to me that the greenhouse problem and the possibility that this will lead to the melting of the two surviving ice caps is sufficiently serious to deal with as a single problem and not complicate it with debates as to how individual countries meet their obligations under any internationally agreed quota system. That is all I want to say by way of introduction to this afternoon's session, which promises to be most interesting.

7
The Greenhouse Effect

KEITH P. SHINE
University of Reading, UK

ABSTRACT

Mankind's activities have, over the past century, led to increased atmospheric concentrations of gases such as carbon dioxide, methane, nitrous oxide and a number of chlorofluorocarbons (CFCs). These gases are termed *greenhouse gases* because their properties are predicted to lead to a warming of the surface. Carbon dioxide is the dominant gas in causing such a warming; however, on a molecule-per-molecule basis the CFCs are very much stronger greenhouse gases and relatively small changes in concentration have been predicted to cause relatively large warmings. The reasons for this strength is outlined in this chapter.

Possible reductions in the emissions of CFCs as a result of the Montreal Protocol may significantly reduce this warming. However, ozone-friendly substitutes may still cause a significant effect.

Possible links between the greenhouse effect and the spring Antarctic ozone depletion are also briefly discussed.

INTRODUCTION

At first sight it may seem a little odd that the greenhouse effect should be discussed at a conference on ozone depletion. Mankind is worried about the depletion of *stratospheric* ozone, whereas it is the effects on *surface* climate that cause so much concern over the increased concentrations of greenhouse gases. Furthermore, there is sometimes confusion in the popular press, and elsewhere, which manages to juxtapose different atmospheric issues such as acid rain, tropospheric ozone increases, stratospheric ozone depletion, the greenhouse effect, etc.

Ozone Depletion: Health and Environmental Consequences
Edited by R. Russell Jones and T. Wigley
© 1989 John Wiley & Sons Ltd

There are, however, a number of reasons why a discussion of the greenhouse effect is important when considering stratospheric ozone. For example, the prediction of changes in atmospheric temperature and circulation as a consequence of increased concentrations of greenhouse gases is important if reliable predictions of changes in stratospheric ozone are to be made (see Chapter 4). There have also been proposals of a link between the greenhouse effect and the Antarctic ozone hole (see Part 6).

This chapter concentrates on a third reason for a linking of the two issues. The chlorofluorocarbons (CFCs), which ultimately lead to ozone destruction, are themselves extremely strong greenhouse gases so that measures aimed at protecting the ozone layer are very relevant to predictions of greenhouse warming. It will be emphasized that ozone-friendly substitutes for CFCs are not necessarily 'greenhouse-friendly'.

Before discussing this subject, this chapter will first outline the basic mechanisms of the greenhouse effect, and predictions from simple models will be shown. The reasons for the efficacy of particular gases for causing a strong greenhouse effect will also be explained.

For a discussion of the analysis of past data in search of a greenhouse 'signal', and more detailed model predictions, see Chapter 8; the impact of increased carbon dioxide and climatic change on ecosystems is considered in Chapter 9.

THE GREENHOUSE EFFECT

The greenhouse effect is not an evil! Without it, life on Earth as we know it could not exist; it keeps the surface of the Earth some 30 °C warmer than it would otherwise be. The concern about the greenhouse effect is that gases we are adding to the atmosphere are increasing that warming effect.

To understand the greenhouse effect we must understand how the Sun's energy is absorbed by the Earth and its atmosphere, and the fate of this absorbed energy.

The Sun emits most of its energy between about 0.2 and 4.0 μm, covering the ultraviolet, visible and near-infrared spectral regions (Figure 1). A very small fraction of that emitted energy is intercepted by the Earth as it orbits the Sun. To balance this constant inflow of energy from space, the Earth must itself emit energy back to space; indeed, satellite measurements do show a very close balance between the energy absorbed and the energy emitted by the planet. Since the Earth is very much colder than the Sun, the bulk of this emission takes place at longer wavelengths than those for incoming solar radiation, specifically between about 4 and 50 μm, a region referred to here as the thermal infrared.

A simple calculation of this energy balance shows that, without an atmosphere, the Earth would have an average surface temperature of about −20 °C; the actual surface temperature is nearer 13 °C.

The reason for this difference stems from the fact that the atmosphere's properties at the shorter solar wavelengths are different from those at the longer thermal infrared wavelengths. Of the solar radiation incident on our planet, about 30% is reflected back to space by clouds, the Earth's surface and the atmospheric gases; another 20% is absorbed by the atmospheric gases (principally by ozone in the ultraviolet and visible, and by water vapour and carbon dioxide in the near infrared) and by water droplets in clouds. The remaining 50% passes through the atmosphere and is absorbed by, and warms, the surface.

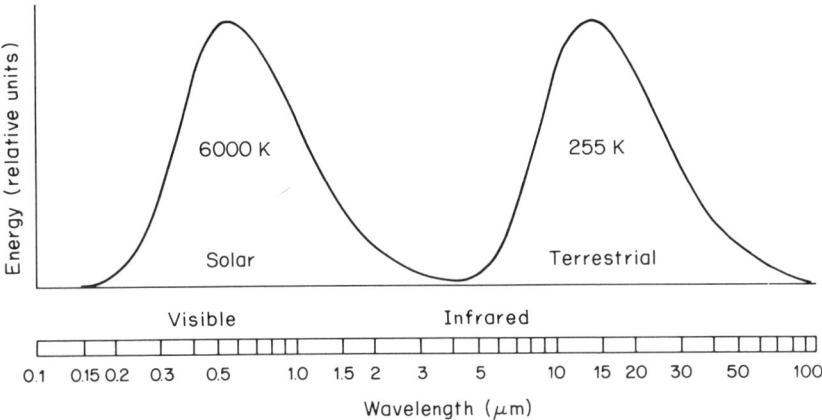

Figure 1   Black-body curves showing the variation of emitted energy with wavelength for temperatures typical of the Sun and the Earth. (Reproduced from SORG[5] by permission of the Controller of Her Majesty's Stationery Office. Crown copyright.)

The energy the surface emits in the thermal infrared 'sees' a rather different atmosphere; clouds, water vapour and carbon dioxide (and, to a lesser extent, a host of other atmospheric constituents) are strong absorbers of radiation at these wavelengths. Consequently, the atmosphere is warmed as much by the absorption of thermal infrared radiation emitted by the surface as it is directly by solar radiation. The crucial point about the greenhouse effect is that the atmosphere itself emits thermal infrared radiation; some of this eventually reaches space, where it helps to balance the incoming solar radiation. Some of it, however, is emitted downwards to warm the surface. Hence the surface is warmed by both solar radiation and thermal infrared radiation (which can be regarded as 'recycled' solar energy). (Whether the greenhouse effect is actually how a greenhouse works is irrelevant to the present discussion!)

Now it can be seen, qualitatively, why the greenhouse effect is of concern. Many gases emitted by mankind's activities are strong absorbers of thermal infrared radiation; their accumulation in the atmosphere makes the atmosphere better able to absorb and emit, so that more energy is emitted down to the surface.

Up to now it has been assumed that all the transfer of energy within the atmosphere is by thermal infrared or solar radiation. This is far from the truth, and it has long been appreciated that a radiation-only view of the atmosphere leads to a surface that is far too warm, and an atmosphere that is far too cold. Such a situation is impossible to maintain; a parcel of air close to the surface is so much warmer (and hence less dense) than the overlying atmosphere that the slightest disturbance would lead to the rising of this buoyant air; such a situation is termed 'unstable'. Atmospheric motions on scale ranging from millimetres to thousands of kilometres act to transfer heat and wipe out any instabilities. A simple analogy is a pan of water heated on an electric ring; the heat is put in at the bottom and the liquid there warms up until it becomes so unstable that it bubbles up towards the surface. In the atmosphere, the motions indicated by cumulus and cumulonimbus clouds transport heat upwards from the surface; such convective motions are most prevalent in the tropics. Less obviously, mid-latitude depressions, so familiar from weather maps, also transport heat both upwards and polewards, so wiping out instabilities. On a globally averaged basis, the energy lost by the surface as a result of the motions is about 1.5 times larger than that lost by radiation.

This region of the atmosphere in which this incessant stirring takes place is called the troposphere and is (normally) characterized by decreasing temperatures with height. It is separated from the overlying stratosphere (where most of the ozone is located) by the tropopause, whose height varies from about 16 km near the equator to about 8 km at the poles. The importance of this in understanding the greenhouse effect is that the surface and troposphere cannot be considered as separate units; they are tightly coupled. When a radiatively active gas is added to the troposphere, it is not the effect that this gas has on the radiative fluxes at the surface that matters, but the change that occurs at the tropopause. Indeed, it is quite feasible (and a good deal of the carbon dioxide forcing happens in this way) that the addition of a gas could cause no change, *initially*, to the surface radiation budget, but still lead to a warming due to changes in the balance at the tropopause.[1] The consequence of the tight coupling between surface and troposphere is that a change at the tropopause is communicated to the surface where a change in the surface energy balance results in a warming.

Our model for the greenhouse effect is now a little less straightforward. The surface is heated by solar radiation and loses that heat both by emitting thermal infrared radiation and by atmospheric motions. The atmosphere emits thermal infrared radiation, both downwards to the surface and upwards to space. The net effect of these emissions is to cool the troposphere; this constant cooling is balanced by the heat transported from the surface by the atmospheric motions, which act to eradicate any instabilities.

The thermal infrared radiation emitted by the surface and the troposphere balances the net incoming solar radiation at the tropopause. For present purposes it can be imagined that the infrared energy is emitted from somewhere in the mid-troposphere. When an extra greenhouse gas is added to the atmosphere, the infrared opacity of the troposphere increases. Energy from the mid-troposphere is now less able to reach space. Consequently, the effective level of emission is from a higher and therefore colder region. Because it is colder, the emitted energy no longer balances the incoming radiation, and the surface and the troposphere warm up until this balance is re-established.

To understand why CFCs are such effective greenhouse gases one further piece of information is necessary. Up to now, it has merely been asserted that atmospheric gases and clouds absorb radiation in the thermal infrared. The manner in which they do this is quite complicated and varies with wavelength. Figure 2 shows the regions of the spectrum at which various gases absorb. For example, water vapour absorbs in two strong bands, one around 6.3 μm, the other at wavelengths longer than 12 μm. Carbon dioxide has a strong band between 12 and 18 μm. Ozone absorbs strongly in a narrow band around 9.6 μm. Except for this ozone band, the region between about 8 and 13 μm has comparatively weak absorption (except when clouds are present) and is termed an 'atmospheric window'.

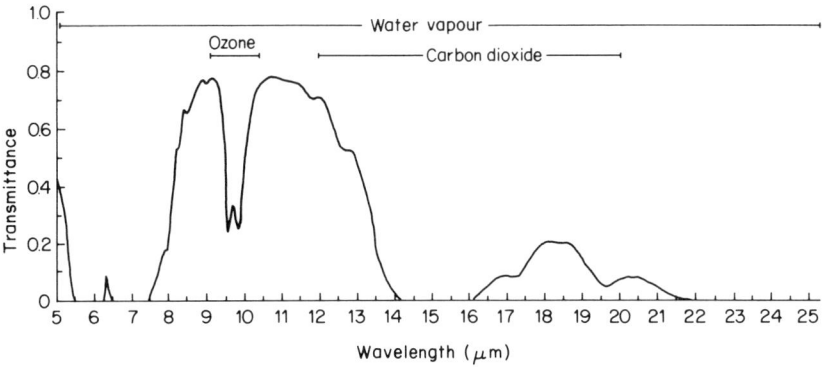

Figure 2   Variation with wavelength of atmospheric transmittance between the surface and space in the thermal infrared for a cloud-free atmosphere in summer at about 60°N. (Reproduced from Selby and McClatchey[14] with permission.)

If we 'sit' at the tropopause and look downwards at different wavelengths, the distance we 'see' depends on the absorption going on. At around 15 μm, both water vapour and carbon dioxide absorb, and we would only see a kilometre or so. In the window region, by contrast, on a cloudless day we could see right down into the lower troposphere. Hence, the infrared energy that reaches the tropopause emanates from many levels; at 15 μm it comes from the cool upper troposphere, and between 8 and 13 μm most of the radiation comes from the warmer lower troposphere. If gases are added to the atmosphere that absorb in the window, and therefore block radiation from the warmer lower troposphere, they will have, potentially, a greater warming effect than a gas that absorbs at 15 μm. A number of gases generated by mankind's activities have absorption bands in the window, as is shown in Figure 3.

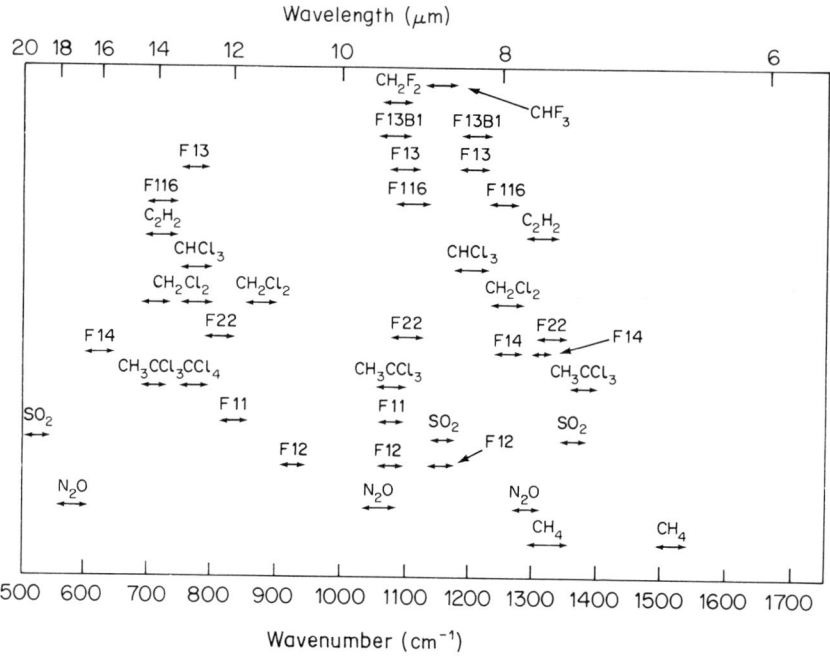

Figure 3 The wavelength of absorption of various trace gases in the atmospheric 'window'. The concentration of many of these gases is increasing due to mankind's activities. (Amended from Ramanathan et al.[1])

The discussion up to now has centred on the troposphere. In the stratosphere the effect of increased concentrations of greenhouse gases leads to a cooling. The explanation for this lies in the way the stratosphere is heated. Instead of being effectively heated from beneath, as the troposphere is, the main heat source in the stratosphere is direct absorption of solar radiation, principally by ozone. This absorption is again balanced by thermal infrared emission. An increased concentration of emitting gases actually enables the stratosphere to emit more effectively, and it is able to balance the absorbed solar radiation by emitting at a lower temperature.

RESPONSE OF THE ATMOSPHERE TO INCREASED GREENHOUSE GASES

Once the atmospheric concentration of a greenhouse gas is increased, the considerations of the previous section lead us to expect the surface and troposphere to warm. The calculation of the size and location of this warming is a difficult task, which requires complex computer models. One of the problems is that the initial warming by the added greenhouse gas can lead to many changes in the climate system that may alter the response. For example, a warmer atmosphere is able to hold more water vapour, which is itself a strong greenhouse gas; this leads to an even greater warming, so that water vapour is said to cause a positive feedback. Similarly, a warmer surface might be expected to have less snow and ice cover. Snow and ice are good reflectors of solar radiation; if their extent is decreased, the amount of solar radiation reflected to space will decrease, causing a further warming. This is another positive feedback.

Other feedbacks are less tractable, and even the sign of the effects are still uncertain. A moister, warmer atmosphere is likely to have an altered cloud amount and the ability of these clouds to absorb and reflect radiation could also change. Current models indicate that the cloud amount will change to increase the warming,[2] whilst for low and middle level clouds the cloud properties will change to reduce the warming.[3] However, both these areas are the subject of much current research and the last word has not yet been spoken. On its own, a doubling of carbon dioxide concentration is predicted to cause a warming of about 1.3 °C, but the effect of all the feedbacks may be to triple this increase.

Climate modellers are only beginning to allow for such factors as changes in vegetation and changes in oceanic circulation, which may cause additional feedbacks of uncertain size and sign.

A further complication in making predictions of the greenhouse warming is that the enormous heat capacity of the ocean acts to slow down the rate of warming. Many models consider only what is called the 'equilibrium warming' due to the increased concentrations of greenhouse gases — that is, the warming that would eventually result after some long time (which may be many decades). To calculate the warming at any given time, a calculation has to be performed whereby the greenhouse gases are increased gradually with time, and the buffering effect of the ocean is included; this gives the transient warming. This is discussed in more detail by Wigley (Chapter 8). Only very recently have such effects been reported using the most complex model type, the three-dimensional general circulation model.[4]

A number of recent calculations have assessed the contribution of gases other than carbon dioxide to the greenhouse effect.[1,4,5] The general conclusion of such works is that carbon dioxide is the gas whose increased concentration will cause the greatest warming, and this will remain so for the foreseeable future. However, the joint contribution from the other gases is far from negligible. Indeed, their joint contribution over recent decades may already be as large as that due to carbon dioxide alone. The principal gases are CFC 11, CFC 12, nitrous oxide and methane, with very uncertain, but possibly significant, contributions from increases in tropospheric ozone and stratospheric water vapour. Table 1 shows the equilibrium warmings due to predicted increases in a number of trace gases by the year 2030, from the results of Ramanathan et al.[6]

Table 1 Predictions of the surface temperature increases resulting from increases in carbon dioxide and other greenhouse gases between 1980 and 2030 by Ramanathan et al.;[6] the increases in CFC concentrations are those projected before the signing of the Montreal Protocol

Gas	Concentration change between 1980 and 2030	Change in equilibrium surface temperature (K)	Percent of CO_2 warming	Accumulated change with respect to CO_2 (%)
Carbon dioxide	111 ppmv	0.7	100	100
Nitrous oxide	75 ppbv	0.1	14	114
Methane	690 ppbv	0.14	20	134
Tropospheric ozone	12.5%	0.06	9	143
CFC 11	0.9 ppbv	0.12	15	158
CFC 12	1.5 ppbv	0.24	31	189
HCFC 22	0.8 ppbv	0.04	6	195
Trichloroethane	1.4 ppbv	0.02	3	198

It must be stressed that such calculations are the subject of large uncertainties, beyond those uncertainties due to size and sign of the feedbacks discussed earlier. The strength of the infrared absorption is, for many of the gases, not particularly well known, and subject to revision, and it is extremely difficult to predict future concentrations of these gases. Not only are the main sources and sinks of some of them poorly understood, but future emissions are subject to political, economic and technological developments.

For the increases in concentrations given by Ramanathan et al.,[6] more recent developments indicate that the effects of some of the CFCs may have been underestimated. For example, measurements of the infrared strength of HCFC 22 by Varanasi and Chudamani[7] indicate that the values used by Ramanathan et al. may have been a factor of 2.5 too small. Thus the equilibrium warming for this gas given in Table 1 may be nearer 0.1 K. Another example concerns CFC 113, much used in the electronics industry. The strength of its infrared bands have only recently been measured (by Varanasi and Chudamani[7] and Rogers and Stephens,[8] whose values disagree by about 25%). On the basis of this strength, the current emission rates and atmospheric lifetimes (about 140 kt/year and 90 years, respectively[9]), a 3% per year growth in emissions would lead to concentrations of about 0.5 ppbv in about 50 years; this may add a further 0.05 K to the warming shown in Table 1.

Table 1 shows that there is an enormous disparity in the ability of individual gases to cause a warming. For example, an increase of 111 ppmv of carbon dioxide leads to a surface temperature increase of 0.7 K. An increase of CFC 12 of just 1.5 ppbv leads to an increase of 0.24 K. This indicates that CFC 12 is, on a molecule-per-molecule basis, able to effect a warming in excess of 10 000 times greater than that caused by carbon dioxide; such a strong effect is also found for many of the other CFCs. On the same basis, nitrous oxide is about 200 times more powerful than carbon dioxide, and methane is about 30 times more powerful. It is instructive to consider why different gases give such divergent effects. There are three general reasons: the strength of the infrared absorption bands, the wavelength of these bands, and the pre-existing quantities of the gas in question. These will now be considered in turn.

Strength of infrared absorption

The determination of the wavelength that a molecule absorbs and the strength of that absorption is rather complex. In general, thermal infrared radiation is absorbed because it can excite vibrations in the bonds holding the atoms in a molecule together, with the rotation of the molecule also being of importance. Carbon dioxide has a strong absorption band near 4 μm, but, as can be seen from Figure 1, this is at a wavelength at which little energy exchange goes on in the atmosphere. The band at 15 μm, which is responsible for most of the predicted greenhouse warming, is one tenth the strength. For the CFCs, the strongest vibrations between carbon and chlorine and carbon and fluorine occur primarily in the 7.5 to 12 μm spectral region; these vibrations make the infrared absorption bands of many CFCs at least ten times stronger than the carbon dioxide 15 μm band on a molecule-per-molecule basis.

Wavelength of absorption

As discussed above, seen from the tropopause, the level at which radiation emanates, and hence the temperature of that emission, depends on wavelength. The effect of a molecule placed in the upper troposphere will depend on the difference between its temperature and the effective emitting temperature of the troposphere. Thus, a molecule absorbing at 15 μm will only interfere with radiation emitted from the cool upper troposphere, whilst one absorbing in the window will absorb radiation emitted from the warm lower troposphere and emit at temperatures characteristic of the upper troposphere, and will have a more substantial impact.

Pre-existing quantities

Related to this previous point, the pre-existing quantities of the absorbing gas are very important. For carbon dioxide, the pre-industrial quantities of the gas were substantial, and, near the centres of its absorption bands, the atmosphere was already opaque over short distances; the addition of extra carbon dioxide molecules has little impact on the opacity. It is generally stated that the warming effect of carbon dioxide is proportional to the logarithm of its concentration.[1] Conversely, for the CFCs, the concentrations are so low that a doubling of the CFC concentration leads to a doubling of the warming effect.

THE GREENHOUSE EFFECT AND SUBSTITUTES FOR CFCs

The previous section stated that the projections of Ramanathan et al.[1,6] showed that CFCs could increase the warming due to carbon dioxide alone by about 50%. More recently, Wigley[10] has shown that changes in emissions of CFCs consistent with the Montreal Protocol could reduce the warming to between one third and one seventh of this figure. However, Wigley warned that substitutes for CFCs may not be benign from the point of view of greenhouse warming. It is the purpose of this section to carry out a rough calculation of the potential effects of some substitutes; it must be stressed that these are little more than 'back-of-the-envelope' estimates.

One possible substitute for CFC 12 as a refrigerant is HFC 134a; this has the desirable property of posing no (direct) threat to the ozone layer as it contains no chlorine. No measurements of the strength of the infrared absorption bands of HFC 134a are known to me, but the presence of four bonds between carbon and fluorine are likely to ensure that it will possess strong bands between 7.5 and 10 μm (see, for example, Rogers and Stephens[8]). It is likely to have a strength intermediate between two molecules already discussed by Ramanathan et al.;[1] carbon tetrafluoride was found to have a warming effect of 0.06 K/ppbv, whilst FC 116 gave 0.13 K/ppbv. On this basis, a value of 0.1 K/ppbv is chosen for HFC 134a for illustrative purposes, although such a choice is somewhat speculative.

At first sight, this replacement would appear to be a weak greenhouse gas. For constant emissions, the equilibrium concentration of any gas is proportional to its atmospheric lifetime. HFC 134a has a lifetime of 10 years compared with 139 years for CFC 12 (see e.g. SORG[9]). After account is taken for the different molecular weights, the equilibrium mixing ratio of HFC 134a is just 9% of that of CFC 12. After accounting for the differing strengths of infrared bands, HFC 134a has only 5% of the strength of CFC 12 on a molecule-per-molecule basis, or about 6% on a pound-per-pound basis. Rogers and Stephens[8] quote a value given by Du Pont of 10% on a pound-per-pound basis.

However, a somewhat different value is obtained if emission rates are not constant. It can be shown that the ratio of the concentrations of two gases whose emissions are growing at x% per year is given by

$$\frac{(x/100)+(1/\tau_1)}{(x/100)+(1/\tau_2)}$$

where τ is the lifetime of the gas in years. If 3% growth in emissions of CFC 12 were to be replaced by 3% growth in emissions of HFC 134a, then, in the long term, the concentrations of HFC 134a would be 34% of those of CFC 12; over shorter periods (less than about 50 years) this percentage is actually higher. The greenhouse effect of the substitute would be nearer 20%.

It is not obvious that such percentages are the relevant quantity in any case; the greenhouse warming due to a particular substance depends on the *actual* mixing ratio of the gas in question and, if unlimited emissions are allowed, large mixing ratios could be attained. A possibly extreme calculation is to assume that HFC 134a, or a similar substance, replaces all the substances controlled by the Montreal Protocol, and that emissions would be unlimited. The current emission rate[9] of the controlled substances is around 900 kt/year which, allowing for 3% growth, could double in 25 years, and more than quadruple in 50 years. With such emissions of HFC 134a, concentrations could exceed 1.6 ppbv within 50 years. The substitute molecules could cause equilibrium warmings of a few tenths of a degree in 50 years and approach 1 K within a century.

Whilst the substitutes may be less effective greenhouse gases than CFC 11 and CFC 12 (mainly because of their shorter atmospheric lifetimes), such figures put the substitutes in the same league as methane and nitrous oxide. It should also be recalled that HCFC 22, itself a possible substitute for the less ozone-friendly species, may have a substantially larger greenhouse effect than has been shown by previous assessments.

The essential point of this discussion is that although the phasing out of such species as CFCs 11, 12, 113, 114 and 115 (if, indeed, such a step is agreed in the future) will reduce the risks to the ozone layer, the simple calculations presented here indicate that the reductions in greenhouse warming may not be substantial if emissions of the replacements are uncontrolled. It is clearly important that the potential greenhouse effects of replacement substances should be examined in greater detail than could be performed here. Nevertheless, in considerations of controls aimed at reducing greenhouse warming, it must always be remembered that carbon dioxide is, and is likely to remain, the gas that contributes by far the largest effect.

THE GREENHOUSE EFFECT AND THE ANTARCTIC OZONE HOLE

As has already been discussed, it is anticipated that the increased concentrations of greenhouse gases will lead to a cooling of the stratosphere. There is some concern that this cooling could cause an intensification of the ozone hole (and, indeed, it may already have had some impact). Eckman and Pyle (Chapter 4) have stressed the important role that polar stratospheric clouds (PSCs) play in the mechanisms believed to cause the depletion of ozone. Most present theories have assumed that the occurrence of PSCs has remained constant whilst the stratospheric chlorine content has increased. However, PSC formation is strongly dependent on temperature, and any lowering of the winter temperature over the polar regions may increase the frequency and areal extent of the clouds, and, potentially, lead to an intensified depletion. This reasoning also presumes that the stratospheric water vapour content remains steady, since this is the quantity that determines the temperature at which the clouds form. This may not be correct, and Blake and Rowland[11] have suggested that increased emissions of methane may already have led to increased stratospheric water vapour content. Future increases would parallel methane increases and this would lead to PSCs being able to form at higher temperatures. In an atmosphere that was itself cooling, the result would be to increase the frequency of PSCs even more.

Little quantitative work has yet been performed on the possible effects of increased frequency of PSCs on ozone depletion. However, on the basis of observations by Karoly,[12] Shine[13] has shown that changes of 1 K (assuming constant stratospheric water vapour content) may lead to changes in areal extent of PSCs by 25%. General circulation models (e.g. Schlesinger and Mitchell[2]) project changes of between 1 and 3 K in polar lower stratospheric temperature for a doubling of carbon dioxide.

CONCLUSIONS

The chlorofluorocarbons possess strong absorption bands in the thermal infrared, which make them extremely effective greenhouse gases. On a molecule-per-molecule basis they are at least 10 000 times stronger than carbon dioxide. Hence, although predicted increases in concentration may be modest compared with the large increases in carbon dioxide concentration, Ramanathan *et al.* predicted that they would cause around 50% more warming than carbon dioxide alone. More recently, Wigley[10] has shown that decreased emissions of CFCs consistent with the Montreal Protocol may reduce this warming to between one third and one seventh of the previous value. However, it has been shown here that unrestricted emissions of substitutes to CFCs may contribute their own significant warming. The need for further detailed measurements and calculations is clear.

It is probably unbalanced to place too much weight on the potential greenhouse effects of CFC substitutes unless measures are also taken to restrict emissions of other greenhouse gases, and in particular carbon dioxide. The impact of the substitutes on reducing the depletion of ozone will be of primary importance until then.

ACKNOWLEDGEMENTS

The Central Electricity Generating Board are thanked for their support of the author through a Research Fellowship. John Thuburn is thanked for his help in calculating atmospheric concentrations of molecules.

REFERENCES

1. V. Ramanathan, L. Callis, R. Cess et al., Climate–chemical interactions and effects of changing atmospheric trace gases. *Rev. Geophys.*, **25**, 1441–1482 (1987).
2. M.E. Schlesinger and J.F.B. Mitchell, Climate model simulations of the equilibrium climate response to increased carbon dioxide. *Rev. Geophys.*, **25**, 760–798 (1987).
3. R.C.J. Somerville and L.A. Remer, Cloud optical thickness feedbacks in the carbon dioxide–climate problem. *J. Geophys. Res.*, **89**, 9668–9672 (1984).
4. J. Hansen, I. Fung, A. Lacis et al., Global climate changes as forecast by Goddard Institute for Space Studies Three-Dimensional Model. *J. Geophys. Res.*, **93**, 9341–9364 (1988).
5. SORG, *Stratospheric Ozone*. Report of the UK Stratospheric Ozone Review Group, HMSO, London (1987).
6. V. Ramanathan, R.J. Cicerone, H.B. Singh and J.T. Kiehl, Trace gas trends and their potential role in climate change. *J. Geophys. Res.*, **90**, 5547–5566 (1985).
7. P. Varanasi and S. Chudamani, Infrared intensities of some chlorofluorocarbons capable of perturbing the global climate. *J. Geophys. Res.*, **93**, 1666–1668 (1988).
8. J.D. Rogers and J.D. Stephens, Absolute infrared intensities for F-113 and F-114 and an assessment of their greenhouse warming potential relative to other chlorofluorocarbons. *J. Geophys. Res.*, **93**, 2423–2428 (1988).
9. SORG, *Stratospheric Ozone 1988*. Second Report of the UK Stratospheric Ozone Review Group, London, HMSO (1988).
10. T.M.L. Wigley, Future CFC concentrations under the Montreal Protocol and their greenhouse-effect implications. *Nature*, **335**, 333–335 (1988).
11. D.R. Blake and F.S. Rowland, Continuing worldwide increases in tropospheric methane, 1978–1987. *Science*, **239**, 1129–1131 (1988).
12. D.J. Karoly, Southern hemisphere temperature trends: a possible greenhouse effect? *Geophys. Res. Lett.*, **14**, 1139–1141 (1987).
13. K.P. Shine, Comment on 'Southern hemisphere temperature trends: a possible greenhouse effect?' *Geophys. Res. Lett.*, **15**, 843–844 (1988).
14. J.E.A. Selby and R.A. McClatchey, *Atmospheric Transmittance from 0.25 to 28.5 μm: Computer Code LOWTRAN 3*. Environmental Research Paper 513. Air Force Cambridge Research Laboratory, Hanson, Mass. (1975).

8
Measurement and Prediction of Global Warming

Tom M. L. Wigley
University of East Anglia, UK

ABSTRACT

Changes in global-mean temperature over the past 100 years are described and interpreted, and possible future changes resulting from increasing concentrations of greenhouse gases are estimated using a simple climate model. Near-surface temperature data from land and marine areas show an overall warming trend of about 0.5 °C over the past 100 years. Superimposed on this trend are variations on inter-annual, inter-decadal and longer time scales. In interpreting the record, one view is that the century time scale trend represents the effect of external forcing factors (such as the greenhouse effect) while the shorter time scale variations represent the 'noise' of natural climatic variability. The observed temperature increase is compatible with model estimates of the amount of warming due to increasing greenhouse gas concentrations, but other explanations are possible. Future greenhouse gas projections are summarized in terms of their combined radiative forcing effects. In doing so, the influence of the Montreal Protocol for limiting CFC production is considered. Future forcing estimates are then used as input to an energy-balance climate model in order to estimate future changes in global-mean temperature.

Ozone Depletion: Health and Environmental Consequences
Edited by R. Russell Jones and T. Wigley
© 1989 John Wiley & Sons Ltd

INTRODUCTION

Changes in the Earth's climate between now and the middle of the twenty-first century are likely to be dominated by the greenhouse effect caused by increasing concentrations of carbon dioxide, methane, nitrous oxide, ozone and chlorofluorocarbons. These greenhouse gases individually and collectively change the radiative balance of the atmosphere, trapping more heat near the Earth's surface and causing a rise in global-mean surface air temperature. Future global warming is virtually certain and likely to be substantial, perhaps as much as 3 °C by the year 2050. This change is similar to the temperature difference between the last Ice Age and today. Changes in all climate variables (precipitation, winds, etc.) will occur in parallel with the global-mean warming, with changes varying substantially from region to region. The details of these regional changes, however, are highly uncertain, and cannot yet be quantified.

Although future changes in climate will be dominated by the effects of man, the climate system has always been a changeable entity. Even over the past century there have been noticeable changes, which provide a backdrop to and a context for an evaluation of the future. This chapter begins, therefore, with a brief review of past climatic change. It is shown that the globe has warmed by about 0.5 °C over the past century. The question of whether this warming can be attributed to the greenhouse effect is then addressed. Future changes in global-mean temperature are then considered, highlighting the important distinction between equilibrium and transient changes.

OBSERVED GLOBAL-MEAN TEMPERATURE CHANGES

Future changes in climate can be set in context by a brief description of changes that have occurred over the past 100 years. Both regionally and in terms of global-mean values, climatic conditions have fluctuated noticeably from year to year, decade to decade and on longer time scales. Only the largest spatial scales will be considered here.

Changes in global-mean temperature can be estimated by combining meteorological observations from the land and ocean areas of the Earth, the bulk of which were originally obtained as routine data for weather forecasting purposes. Many hundreds of millions of observations go into producing area-average figures spanning the last 100 years or so. Early estimates (reviewed by Wigley et al.[1]) were based solely on land data. Before they can be used, these data must be critically examined for inhomogeneities (i.e. for variations arising from non-climatic sources, such as changes in station location or observing times, and effects such as urban warming). Jones et al.[2,3] have produced the most comprehensive analysis. Other compilations[4,5] do, however, give similar results. This would be expected because all analyses have common data sources.

More recently, the vast amounts of information collected by ships at sea have been compiled by groups in the UK and USA.[6,7] These data have been combined with the land data to give a more reliable picture of global-mean changes.[8] The marine data are particularly important, because they represent some 70% of the Earth's surface area (although coverage of this area is incomplete, simply because it is limited to the regions where ships travel). Including marine data into the estimates is, however, not a simple task because these data have, over the years, been collected by a variety of means. The most commonly used data are sea surface temperatures, but marine air temperatures are also used, generally giving consistent results. For sea surface temperatures, adjustments must be made to the data to account for the changes in measurement technique: from the use of wooden buckets in the nineteenth century, to canvas buckets in the early twentieth century (which suffer from evaporative cooling), to engine cooling water intake observations in recent decades. Different methods have been used to make these adjustments.[6,8] For the twentieth century, the methods give similar results. For the nineteenth century, however, there are still discrepancies which have yet to be adequately resolved: some analyses of marine data (those of Folland et al.[6]) give temperatures up to 0.2 °C warmer than the data of Jones et al. used here (Figure 1).

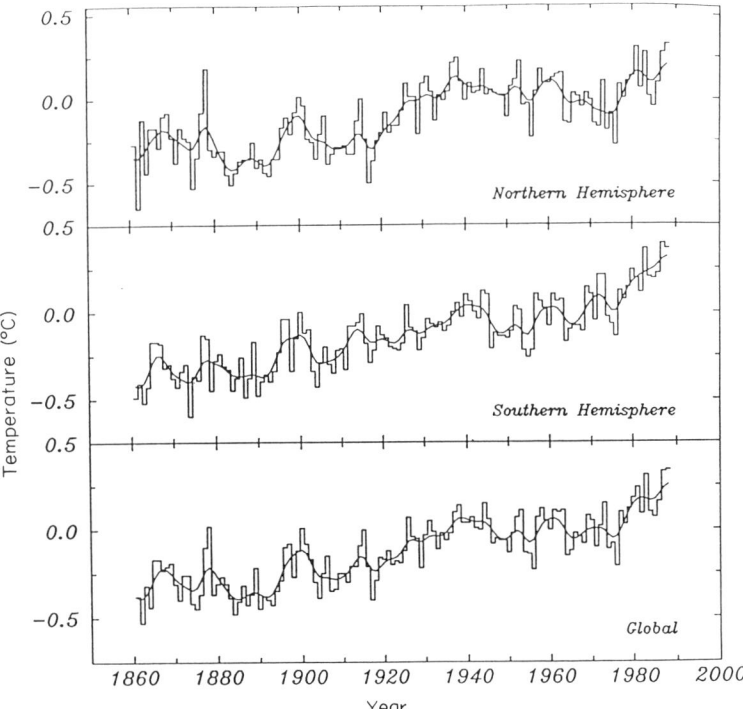

Figure 1 Hemispheric- and global-mean temperature changes based on land and marine data (from Jones et al.,[8] updated). Changes are relative to a reference period of 1950–79. For the most recent years, marine values were provided by the UK Meteorological Office.

These analyses show that the near-surface air temperature averaged over the globe has increased by about 0.5 °C since the late nineteenth century. Figure 1 illustrates this by showing hemispheric- and global-mean changes. Parallel changes in the temperature of the lower troposphere have also occurred (see Wigley et al.[1] for a recent review), but these data extend back only to the late 1950s.

As is evident in Figure 1, the recent global-scale warming has not been a continuous upward trend. Nor has the warming been spatially homogeneous, and trends have varied substantially from region to region. These spatial differences are illustrated in Figure 2, which shows annual-mean temperature trends over the Northern and Southern Hemispheres for 1947–86 (from Jones et al.[9]). Over the past 40 years, much of the North Atlantic and western Europe has undergone a noticeable cooling.

This cooling, and indeed the cooling that is evident in the Northern Hemisphere mean time series shown in Figure 1 between 1940 and 1975, appears to be inconsistent with the greenhouse hypothesis. Although the cooling is undoubtedly real, it cannot be taken as evidence that there is no greenhouse effect, or even that the magnitude of the greenhouse effect is small. Rather, it is a graphic illustration of the extent and magnitude of natural climatic variability, the 'noise' against which the greenhouse 'signal' must be detected, and upon which future greenhouse warming will be superimposed. These issues will be discussed further below.

FUTURE GLOBAL-MEAN TEMPERATURE CHANGES

Climate changes in the future will depend on how the climate system is forced by external factors and how it varies naturally due to internal interactions between the various components of the system. Some of the potential external forcing factors contribute to the climate's natural variability: examples are explosive volcanic eruptions and changes in solar irradiance. Other factors may be considered either internal, external or both, depending on one's point (or scale) of view. For example, changes in albedo (i.e. reflectivity) may occur as a result of changes in cloudiness or changes in sea ice and/or snow cover. Although such changes occur naturally because of internal climate processes, they may also be considered in some situations as external factors. Recent changes in greenhouse gas concentrations are clear external factors, but qualitatively similar changes

Figure 2 Total changes in annual-mean temperature (°C) over the period 1947–86, illustrating the spatial variability of changes in climate. The changes shown were calculated by fitting a linear trend line to the data series at each grid point on a 5° latitude by 10° longitude grid. For the Antarctic region the linear trend was fitted to annual data for the 1957–86 period only, since there are no data prior to 1957. Shading indicates areas where there are insufficient data (or no data) for a trend to be calculated over the study period.

MEASUREMENT AND PREDICTION OF GLOBAL WARMING

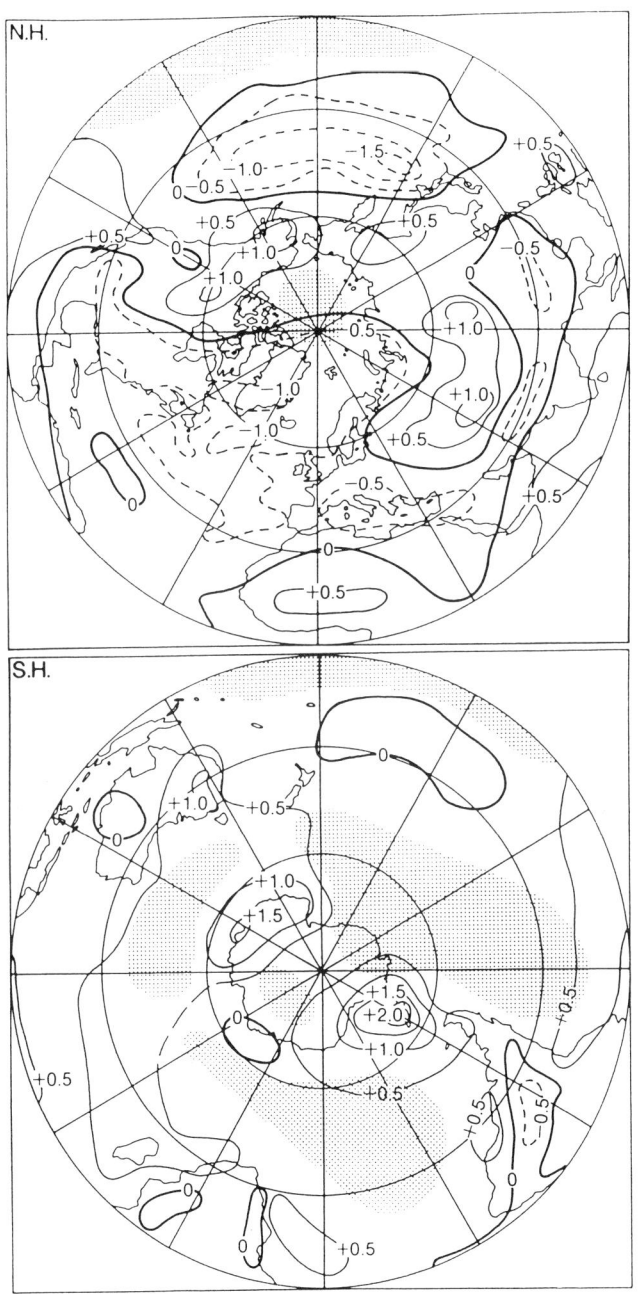

have occurred in the past as a result of natural processes. During the last ice age, for example, carbon dioxide and methane levels were appreciably less than the pre-industrial levels prevailing in the eighteenth century.

Of all these factors, only one has a well defined past history that can be projected forward into the future; namely, the combined effect of greenhouse gas concentration changes. (Minor solar irradiance changes that occur in phase with the sunspot cycle can also be predicted to occur, but these are insignificant compared with the greenhouse effect.)

Three things will determine the magnitude of future greenhouse-gas-induced global-mean warming: the future concentrations of the gases; the equilibrium sensitivity of the climate system to external forcing; and delays in the response due to the thermal inertia of the oceans. The changes in temperature that will actually be observed, however, will be the sum of greenhouse forcing effects and natural variability. Natural variability will always contribute an important, and largely unpredictable, component to climatic change, although in future decades the greenhouse effect is likely to play the dominant role.

Before putting numerical values to future climatic change, there is an important theoretical distinction that must be made; namely, that between the equilibrium and the transient response. Suppose that some external forcing factor is imposed instantaneously, such as a 1% increase in solar irradiance. How will the climate system respond? The response will certainly not be instantaneous; the global-mean temperature would begin by increasing quite rapidly and then less and less rapidly until it finally reached a value arbitrarily close to a new steady state. The response is similar in principle to the way a car responds to pressure on the accelerator. With a vehicle, the slow response is due to the fact that the car has a finite weight (or mass inertia), while with the climate system the slow response is due to the oceans' *thermal* inertia. The same situation applies even if pressure on the accelerator is gradually increased; the car's response lags behind that corresponding to the pedal's position, so that, if the pedal stops at a certain point, the car will still accelerate for a while before reaching a constant velocity (or steady state) corresponding to the pedal's stopping point.

In the climate system, the analogue of the steadily increasing pedal pressure is the steadily increasing forcing due to greenhouse gas concentration build-up. The time-varying temperature changes are referred to as the *transient* response. The final temperature level that would be attained for a given forcing level (equivalent to the final speed of the car for a given pedal position) is called the *equilibrium* response. Just as in a car, the lag between transient and equilibrium response depends on a variety of factors: the weight of the car (i.e. the oceans' effective heat capacity); the rapidity of the increase in pedal pressure (i.e. the rate of change of greenhouse gas concentrations); and the power of the car (i.e. the climate sensitivity).

Climate sensitivity in the present context is usually judged by the equilibrium global-mean temperature change that would eventually occur if the carbon dioxide level were doubled, denoted by ΔT_{2x}. This must be determined using some form of climate model. Its value depends critically on a variety of feedback processes that exist within the climate system. An example is water vapour feedback. If we warm the world, more water will evaporate from the oceans, adding water vapour to the atmosphere. Since water vapour is a greenhouse gas, this will increase the greenhouse effect and cause further warming. Numerous other feedback mechanisms exist involving clouds, sea ice, snow cover, the ocean circulation, etc. Since their quantitative effects are uncertain, the magnitude of ΔT_{2x} is also uncertain. The best estimate given in recent reviews of the greenhouse problem is that it lies within the range 1.5–4.5 °C with about 95% confidence.[10,11] Some models, however, have given values noticeably higher than the upper limit of this range.[12]

In order to estimate future global-mean temperature changes, we must consider the transient response and somehow account for the damping effect of oceanic thermal inertia. A variety of models is potentially available for this purpose but, to date, only relatively simple models have been used in any systematic way. These simple models describe vertical heat transport in the oceans as a diffusion process, often including, in a simplified way, the effects of vertical advection (for a review, see Hoffert and Flannery[13]). The results described below are based on the model of this type developed by Wigley and Raper.[14] The main model parameters that determine its response are the climate sensitivity and, to a lesser degree, the vertical diffusion coefficient. As noted above, climate sensitivity determines not only the equilibrium response, but how rapidly equilibrium is approached. For any given forcing, the damping or lag is greater for higher climate sensitivity.

Future warming must, of course, depend on future forcing; i.e. on future greenhouse gas concentrations. These are subject to considerable uncertainty. The present (1988) level of the most important greenhouse gas, carbon dioxide, is just over 350 ppmv. This is some 25% more than the pre-industrial (late eighteenth century) level, which was around 279 ppmv. In projecting future changes, the overall effect of changes in carbon dioxide, methane, CFCs, etc., is often expressed in terms of an *equivalent* carbon dioxide concentration. Using a base level of 279 ppmv in the late eighteenth century, the present equivalent carbon dioxide level is just over 400 ppmv. This is some 50 ppmv above the actual carbon dioxide level, indicating that the other greenhouse gases have, in terms of their radiative forcing, effectively added 50 ppmv to the carbon dioxide level. The overall radiative forcing change since the late 1700s is more than 2 W/m^2, equivalent to increasing solar irradiance by about 1%.

For the future, the best estimate is that the equivalent carbon dioxide level will reach 558 ppmv (double the pre-industrial level) by the late 2020s (see Wigley[15,16]). Projected changes, together with an estimate of the uncertainty, are shown in Figure 3.

In Figure 3, the projected changes have been made after consideration of the Montreal Protocol for reducing CFC production. CFCs are estimated to have contributed approximately 20% to total greenhouse forcing over the period 1950–85 (compared with 22% due to methane, 6% due to nitrous oxide and 52% due to carbon dioxide), and, before the Montreal Protocol, they were expected to

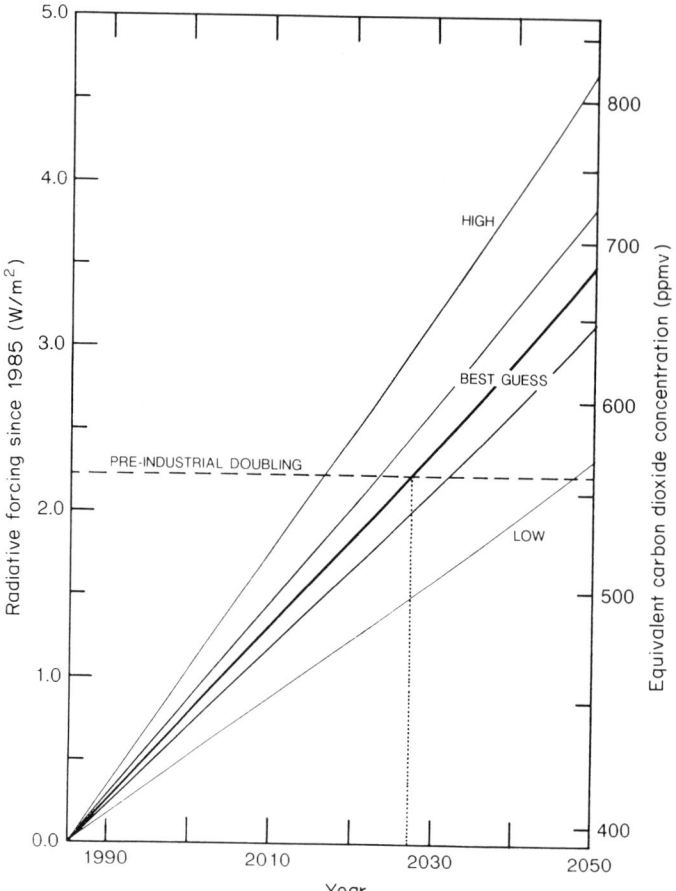

Figure 3 Projected greenhouse forcing, 1985–2050. The values are expressed in terms of radiative forcing and the equivalent CO_2 concentration. To give some idea of the magnitude of the radiative changes, a 1% increase in solar irradiance corresponds to about 2.4 W/m^2.

contribute even more in percentage terms in the future. However, if the Protocol is rigidly adhered to, if all countries (including India and China) eventually sign the Protocol, and if the level of substitution by other related chemicals is relatively small, the future greenhouse contribution from CFCs will be much less. This is an optimistic viewpoint, especially since some proposed substitutes are strong greenhouse gases (see Chapter 7), but the Protocol will probably be strengthened in the future so the net effect on greenhouse forcing is likely to be one of considerable benefit.

Past and projected forcing can be used as input to a climate model to estimate global-mean temperature changes. In Figure 4, the forcing used is as observed up to 1988, together with the best estimate of future forcing from Figure 3 subsequently. Figure 4 also shows the observed global-mean temperature changes as given in Figure 1. The only uncertainty accounted for in Figure 4 is that due to uncertainty in the climate sensitivity, which produces a band of model-based estimates of global warming. Other model uncertainties would make this band a little wider.

It can be seen immediately that the observations are consistent with the hypothesis of a long-term, greenhouse-gas-induced warming. However, it is not possible to say whether the observations fit the lower curve (low sensitivity) or upper curve (high sensitivity) better; the choice would depend on whether one concentrates on the earlier or later part of the record.

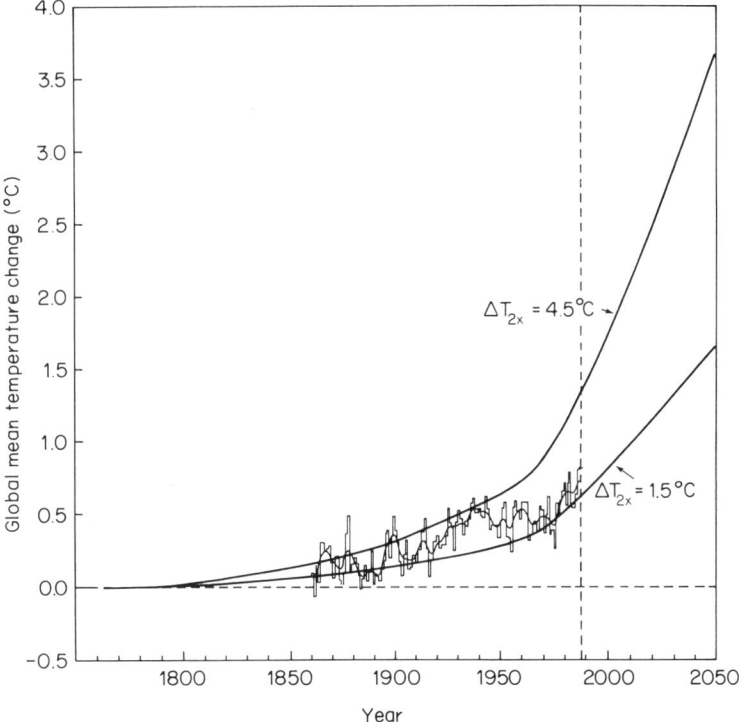

Figure 4 Comparison of observed and modelled global-mean temperature changes. The observed data are the same as in Figure 1, lowest panel. The modelled data (smooth curves) correspond to observed forcing prior to 1986 and 'best guess' projected forcing after that. The two model curves show results for extreme values of the climate sensitivity, namely equilibrium $2 \times CO_2$ warmings of 1.5 °C and 4.5 °C. The model runs begin in 1765.

In spite of the overall consistency, there are important aspects of the observed record which differ from model expectations. The most obvious is the lack of observed change between 1940 and the mid-1970s, which is at odds with the model-predicted warming. If we were certain of the magnitude of the greenhouse effect, these discrepancies would be clear indications of the magnitude of natural climatic variability; either internally generated and/or externally forced variability. Independent calculations show that natural variability of this magnitude is to be expected. However, since we cannot quantify this natural variability, we are unable to quantify the residual greenhouse component of the record. Indeed, it is still possible that *none* of the observed warming is the result of an increased greenhouse effect, although this is unlikely. Most scientists, therefore, say that the climatic consequences of increasing greenhouse gas concentrations have not yet been proved beyond doubt, but they also point out that there is no evidence that the models are seriously wrong. Furthermore, just as it is possible that much of the observed long-term warming trend is a natural phenomenon, it is also possible that there has been a strong greenhouse warming which has been partially offset by natural variability (namely, a long-term natural cooling).

Figure 4 gives projections into the future, but these projections do not account for all possible sources of uncertainty. These are accounted for in Figure 5, which is based on the extremes of forcing shown in Figure 3, together with a range of climate sensitivities (ΔT_{2x} = 1.5–4.5 °C) and a range of ocean diffusivities (K = 0.5–2.0 cm^2/s). Global-mean warming over the next 40 years or so is expected to be between 0.5 and 2.5 °C, with a best estimate of 1.5 °C. Even at the low end of this range, the rate of warming would be almost three times the average rate over the past century. At the high end, the warming rate would exceed anything observed in the past record: it would be about twice as rapid as the warming that has occurred over the past 15 years, but sustained for many decades — perhaps a century or more.

SUMMARY AND CONCLUSIONS

Instrumental observations over the past 100 years or so show that global-mean temperatures have warmed by about 0.5 °C. There is some uncertainty in this figure because some data problems are difficult to overcome, and because, even today, data coverage is not complete. Unresolved problems with the nineteenth-century data mean that it is possible that the global-mean warming has been slightly less than 0.5 °C.

When compared with model-based predictions of the warming due to increasing greenhouse gas concentrations, the observations are consistent with the theory. If the observed century time scale warming were due solely to the greenhouse effect, the implied climate sensitivity would be at the low end of the accepted range; about 2 °C for the equilibrium warming due to doubling the carbon dioxide concentration compared with 1.5–4.5 °C. However, natural variability may well have significantly affected global mean temperatures. Because of this fact, it is still impossible to assert, unequivocally, that the

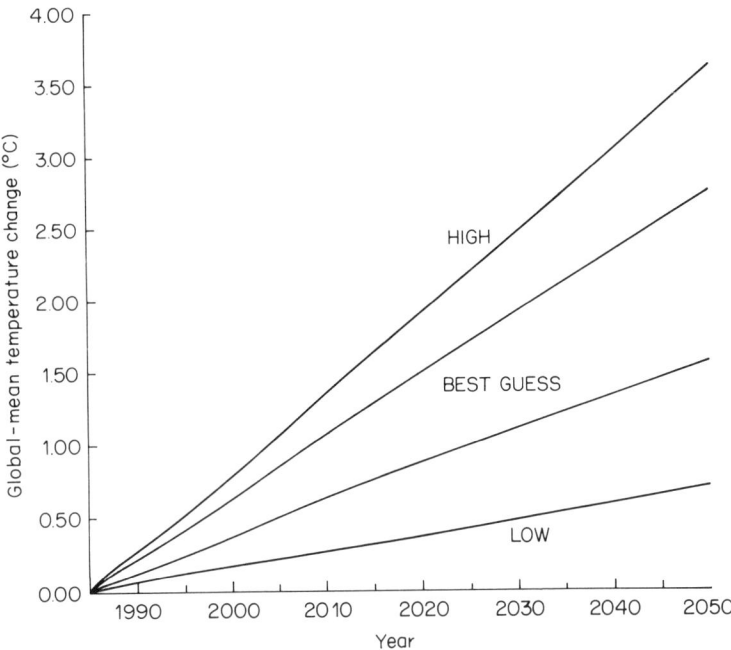

Figure 5 Range of possible global-mean warming, 1985–2050. The low and high extremes correspond roughly to 95% confidence limits accounting for uncertainties in future forcing (see Figure 3), in climate sensitivity (through ΔT_{2x}) and in ocean mixing rates (through the diffusivity, K).

greenhouse effect has been detected, but neither is it possible to preclude quite high values of the climate sensitivity.

Warming projections into the future depend on future forcing, the magnitude of the climate sensitivity, and on the damping effect of oceanic thermal inertia. If uncertainties in these quantities are accounted for, the global-mean warming over the period 1988–2030 is expected to lie in the range 0.5–2.5 °C. Although natural variability will modulate this warming and could even produce periods of global-mean cooling if the smaller estimates of warming rate are correct, on time scales of decades and longer, future climate is likely to be dominated by the greenhouse effect. Rates of change are likely to exceed anything previously experienced on the century time scale by a factor of two to ten.

What can we do about this? Already, some progress has been made through the Montreal Protocol to reduce CFC production. Although introduced primarily to protect the Earth's ozone layer, the Protocol will reduce future greenhouse forcing, although by how much depends on whether or not all countries sign the Protocol, on the extent to which radiatively active substitutes are used to replace controlled substances, and on future modifications to the Protocol. In principle, similar protocols could be introduced to effect reductions in carbon dioxide emissions (e.g. by reducing dependence on fossil fuels for energy production, and by cutting back the current alarming rate of deforestation). However, the sources of carbon dioxide and of other greenhouse gases are so diverse and so tightly bound to population growth and development that even a partial solution to the greenhouse problem will probably require radical measures.

To compound the problem, we may already have committed ourselves to substantial global warming. Because of the damping effect of oceanic thermal inertia, the observed warming at any time must be less than the instantaneous equilibrium value. Thus, if the climate sensitivity were high, the globe would warm by more than 1 °C even if we were able to keep all greenhouse gas levels at their present value. (At the low end of the climate sensitivity range, the current warming commitment is 0.2 °C.) Future planning must take this warming commitment into account.

This paper has concentrated on a single facet of the greenhouse problem; global-mean temperature change. As this variable changes, all other climate variables will change in concert. At the regional level, we can expect large changes in temperature and precipitation, which will vary according to season. At present, however, models are unable to predict the details of future climatic change at the regional level. Sea level will almost certainly rise due to thermal expansion of the oceans and the melting of land-based ice. Between now and 2030, a rise of 17–26 cm is the current best estimate. This is equivalent to a rate of rise some three to five times the rate that has occurred over the past century. As with climate change, global sea level rise would be modulated by local effects due to land movement, and future predictions need to take these factors into account.

The greenhouse effect to date has been masked by natural variability. This is unlikely to be the case in the future, but it may be some time before we can prove its existence beyond doubt and reliably quantify its magnitude. Nevertheless, we cannot afford to wait for absolute confirmation, since future changes are potentially so large that they will be highly disruptive. The lead time required to change energy policies, or to effect other changes that might offset future climatic change, is of the order of decades. When this fact is considered in conjunction with the warming commitment concept, it is clear that action should begin as soon as possible.

ACKNOWLEDGEMENT

This work was supported by the US Department of Energy.

REFERENCES

1. T.M.L. Wigley, J.K. Angell and P.D. Jones, Analysis of the temperature record. In M.C. MacCracken and F.M. Luther, eds, *Detecting the Climatic Effects of Increasing Carbon Dioxide*, US Department of Energy, Carbon Dioxide Research Division, Washington DC, pp. 55–90 (1985).
2. P.D. Jones, S.C.B. Raper, R.S. Bradley, H.F. Diaz, P.M. Kelly and T.M.L. Wigley, Northern Hemisphere surface air temperature variations, 1851–1984. *J. Climate Appl. Meteorol.*, **25**, 161–179 (1986).
3. P.D. Jones, S.C.B. Raper and T.M.L. Wigley, Southern Hemisphere surface air temperature variations, 1851–1984. *J. Climate Appl. Meteorol.*, **25**, 1213–1230 (1986).
4. K.Y. Vinnikov, P.Y. Groisman, K.M. Lugina and A.A. Golubev, Variations in Northern Hemisphere mean surface air temperature over 1841–1985. *Meteorol. Hydrol.*, **1987(1)**, 45–53 (1987) (in Russian).
5. J. Hansen and S. Lebedeff, Global trends of measured surface air temperature. *J. Geophys. Res.*, **92**, 13345–13372 (1987).
6. C.K. Folland, D.E. Parker and F.E. Kates, Worldwide marine temperature fluctuations 1856–1981. *Nature*, **310**, 670–673 (1984).
7. S.D. Woodruff, R.J. Slutz, R.J. Jenne and P.M. Steurer, A comprehensive ocean-atmosphere data set. *Bull. Am. Met. Soc.*, **68**, 1239–1250 (1987).
8. P.D. Jones, T.M.L. Wigley and P.B. Wright, Global temperature variations, 1861–1984. *Nature*, **322**, 430–434 (1986).
9. P.D. Jones, T.M.L. Wigley, C.K. Folland and D.E. Parker, Spatial patterns in recent worldwide temperature trends. *Climate Monitor*, **16**, 175–185 (1988).
10. M.C. MacCracken and F.M. Luther, eds, *Projecting the Climatic Effects of Increasing Carbon Dioxide*, US Department of Energy, Carbon Dioxide Research Division, Washington DC, 381 pp (1985).
11. B. Bolin, B.R. Döös, J. Jäger and R.A. Warrick, eds, *The Greenhouse Effect, Climatic Change, and Ecosystems*, SCOPE Vol. 29, John Wiley & Sons, Chichester, 539 pp (1986).
12. M.E. Schlesinger and J.F.B. Mitchell, Climate model simulations of the equilibrium climatic response to increased carbon dioxide. *Rev. Geophys.*, **25**, 760–798 (1987).
13. M.I. Hoffert and B.P. Flannery, Model projections of the time-dependent response to increasing carbon dioxide. In M.C. MacCracken and F.M. Luther, eds, *Projecting the Climatic Effects of Increasing Carbon Dioxide*, US Department of Energy, Carbon Dioxide Research Division, Washington DC. 149–190 (1985).
14. T.M.L. Wigley and S.C.B. Raper, Thermal expansion of sea water associated with global warming. *Nature*, **330**, 127–131 (1987).
15. T.M.L. Wigley, The effect of model structure on projections of greenhouse-gas-induced climatic change. *Geophys. Res. Lett.*, **14**, 1135–1138 (1987).
16. T.M.L. Wigley, Future CFC concentrations under the Montreal Protocol and their greenhouse-effect implications. *Nature*, **35**, 333–335 (1988).

9
Vulnerable Ecosystems

MICHAEL OPPENHEIMER
Environmental Defense Fund, USA

ABSTRACT

Climatic change due to anticipated greenhouse warming will affect ecosystems through changes in temperature, precipitation, soil moisture, storm frequency and intensity, and sea level. For slowly migrating species, including trees in mid-latitude forests, successful adaptation to expected rates of temperature change may not be possible. Even for gradual rates of temperature change, substantial shifts in forest type and biomass will occur according to analyses of anticipated life zone changes. Human barriers to migration of plants and animals will further inhibit successful adaptation, particularly for species currently restricted to 'island refuges' such as national parks. The continuous nature of projected climatic warming will inhibit attempts to ameliorate these difficulties. Decreased soil moisture in mid-continent areas could reduce habitats for species dependent on wetlands.

Sea level rise will remake coastal ecosystems. Loss of wetlands and barrier beaches in developed areas will occur due to inhibition of inland migration by settlements and other land uses. Reduced river flow due to upstream diversions will combine with higher sea level to disrupt the coastal environment completely in heavily populated deltaic regions, particularly at low latitudes. Increased tropical storm frequency or intensity would also be highly problematic in the humid tropics, while semi-arid regions would be vulnerable to changes that either decrease soil moisture or increase the variability of precipitation.

The stresses on ecosystems due to climatic change cannot be separated from other problems already affecting them. Regional air pollution has significantly disturbed mid-latitude forests. Human population pressures are destroying forests in the humid and semi-arid tropics. Excessive nutrient supply is leading to the gradual eutrophication of coastal seas and estuaries. Climatic change will exacerbate these stresses continuously at an increasing rate. Such interactive effects may already be observable, including synergistic disturbance of ecosystems. One net consequence of these stresses may be a decrease in terrestrial biomass accompanied by an increase in biomass, sedimentation and eutrophication in coastal ocean ecosystems.

INTRODUCTION

I would like to thank the organizers of this conference for inviting me to speak about the consequences of global warming for natural ecosystems, because it is natural ecosystems that will bear the full brunt of climatic change. It has been argued by many experts that agriculture may well adjust to climatic change without disruption of global food supplies. It may be argued that human societies as a whole will adapt, particularly if the change is not too rapid, although not without considerable pain. However, it is difficult to support the notion that rapid global warming will entail anything less than a disaster for many natural systems. It is certainly plausible to argue that climatic change will remake the face of the earth.

I want to discuss some of these consequences in this lecture. This discussion is largely based on the results of the Villach 1987 conference, Developing Policies for Responding to Climatic Change.[1] The papers from this conference will be published as a special issue of *Climatic Change*. However, it should be understood from the outset that, if our quantitative understanding of the potential climate changes is limited, our knowledge of ecological effects is primitive. Given the uncertainty, this discussion is really about vulnerability, as the title indicates, rather than quantitative estimates of effects. Nevertheless, there are three general principles of global warming from which the notion of disastrous change proceeds, even while the specifics remain uncertain. They are as follows:

(1) Climate change will be continuous as long as greenhouse gases are emitted in quantities anywhere near current emissions.
(2) Based on current predictions, the sustained rate of global warming may well exceed both recent and more remote historical rates by a large margin.
(3) Climate change is far from the only stress occurring in a natural world that is already under siege from air and water pollution, deforestation, soil erosion and other human pressures.

I shall return to these points later, while I focus on only two types of ecosystem: mid-latitude forests of the type that cover much of Europe and the USA, and coastal ecosystems such as wetlands and barrier beaches. But remember that the effects of the warming will be global.

Let us set the stage for this discussion by examining once again a projection of future global mean temperature. The recent Villach-Bellagio analysis[1] projected that the global mean temperature would most likely increase at a rate of about 0.3 °C/decade using a business-as-usual emissions scenario, with the warming rate perhaps twice as fast at high latitudes. For comparison, except for short periods, the historical rate of change of global mean temperature may not have exceeded about 0.1 °C/decade either in recent times or during the last glacial retreat.

In other words, in the future, ecosystems may be required to respond much more rapidly than in the past to climatic change.

CONSEQUENCES OF GLOBAL WARMING

Consequences of global warming are imposed through *local* changes in temperature, precipitation and storm patterns. For instance, several models[2] have been used to project local changes in summer season soil moisture, which governs runoff and water availability for plants and animals. In at least one model,[2] summer soil moisture is projected to decrease over much of the globe, particularly the mid-continent areas, over the next 100 years in the business-as-usual scenario. However, there remains controversy over the extent of this drying, with different models giving substantially different results, or even similar results for different reasons.

The clearest way to understand the consequences of these changes for ecosystems is to consider where things live now and where they may possibly find hospitable climate conditions in the future. One simple way to do this is by examining Holdridge life zones[3] derived from temperature and precipitation. Within 100 years, the warming could eliminate the Arctic ecosystem from Alaska and move other zones by about 500 km to the north, according to this analysis. Species restricted to island refuges, such as national parks or mountaintops, may simply have nowhere to shift to and may disappear. Forests along the southern margins of Holdridge zones will shrink, while those at the northern end could expand.

The situation is complicated, and probably further exacerbated by the following:

(1) Ecosystems don't move as a piece; they scatter. So ecosystems won't shift; they will change.
(2) Soils aren't always appropriate at the latitudes that species would be pressed to move toward.

The combined effect of these two factors is projected to bring biomass declines of one third to one half in US southern pine forests, for instance,[4] over the next century.

(3) Trees can migrate only slowly. Trees moved at 20–200 km/century during the glacial retreat, a rate that may indicate the limit of their ability to move. Also, human barriers now intervene. Compare that rate with the 400 km/century derived from the Villach–Bellagio scenario.[1] In other words, many tree species may simply disappear because they cannot compete with the rate of climate zone shift.

There are important uncertainties that require exploration. Will increased carbon dioxide 'fertilize' perennials as it does some crops? Or will weeds develop a selective advantage? Can human intervention through seed-spreading sustain trees or will silviculture itself fail in the face of rapid warming and zone-shifting?

How may future terrestrial systems actually look? Wetland wildlife habitats will dry up where summer soil moisture is decreased. Wildlife is particularly vulnerable to consecutive dry years, and increased dryness will increase the probability of successive dry years. Forests are already declining in Europe and North America, in part due to air pollution. What will happen when a warming stress is added? Synergistic effects can be expected to accelerate the current decline, and the forest fires of this summer in the USA remind us that forests do not die quietly, particularly in a dry world.

Now let's examine another type of ecosystem: that of the coastal environment. Here we have the wetlands and estuaries which are the homes for migrating birds and the nurseries for half the fish caught for food. Here we have beaches. Half of humans now live near coasts and this fraction is increasing. In Asia, they often live in low-lying, heavily cultivated deltaic regions, particularly vulnerable to flooding. In the coastal zone, fresh drinking water and salt water lie close together. The coastal zone is also the receiver of all upstream changes, such as those that might occur in nutrient and sediment flows due to forest decline. This coastal environment will be exposed to the combined effects of a number of stresses; higher temperature and sea level, higher carbon dioxide concentrations, and higher nutrient flows. Coastal ecosystems may be, in short, remade and destroyed. Sea level may rise much faster than the recent historical rate due to expansion of ocean water and melting of land ice. Societal adaptation by movement or protection of infrastructure and beaches will be costly. In deltaic areas such as Bangladesh, resources for adaptation are limited, so abandonment and increasing loss of life can be expected.

Once again, it is the natural ecosystems that simply will not adjust successfully. Wetland migration is unlikely. A one third to two thirds loss in US north-east coast wetlands has been projected.[5] Areas like the Baltic and North Sea, already under stress due to a nutrient excess, will be further damaged by algal blooms encouraged by increased temperature, carbon dioxide and nutrients. At the very least, species composition changes will occur. Outside the mid-latitudes there are other vulnerabilities. Low-latitude vegetation is particularly vulnerable to climate variability where deforestation and soil erosion are already a problem. Climate feedbacks involving ice and tundra at high-latitude regions could accelerate the warming and create additional stresses for ecosystems.

Is there any good news? Some ecosystems might benefit (if any change is beneficial) as measured by biomass increase. The northern boreal forests may provide an example, if they can move fast enough and if the warming stops at some point.

CONCLUSION: THE NEED FOR EMISSIONS ABATEMENT

Let me summarize by putting these changes in context. Ecological change has occurred before. Recovery from the last glacial age is still occurring for some species. Cyclic stresses like fires are important and natural.

However, global warming will probably occur faster and be more sustained than previous climate changes and will not stop to allow stabilization or adjustment. We are moving into a world of continuous change in which slow-moving species will drop by the wayside and entire ecosystems will disappear. Uncertainties are very large, but because consequences lag far behind the emissions responsible we cannot wait to observe signs of ecological decline and then pull back. If we don't like the consequences of warming after the fact, our recourse is limited since the changes are, in a practical sense, irreversible on a human time scale.

The biosphere is also a storehouse of potential greenhouse gases — carbon dioxide, nitrous oxide and methane. The warming and associated disturbance of the biosphere could well accelerate the growth of greenhouse gases in the atmosphere and, thus, cause further increases in the rate of warming.

Finally, let me finish on an optimistic note. These changes are not inevitable and the future remains, by and large, a matter of choice.

Two sorts of emissions 'futures' are imaginable: one, business as usual; another, a course of emissions, determined by substitution of efficiency measures and renewable energy for current use patterns, that would slow global warming to a pace to which ecosystems might adapt and which may ultimately stabilize the climate. Achieving that second course is a daunting challenge, but I hope everyone will keep in mind that human beings and societies are ultimately dependent on the biosphere. And as ecosystems go, so ultimately goes the human race.

REFERENCES

1. J. Jager, *Developing Policies For Responding To Climatic Change,* Proceedings of conferences at Villach, Austria, 28 September to 2 October 1987 and Bellagio, Italy,

9-13 November 1987. World Meteorological Organization and United Nations Environment Programme, 1-53 (1988).
2. J.F.B. Mitchell, The greenhouse effect and climate change. *Rev. Geophys.*, **27**, 115-139 (1989)
3. W.R. Emanuel, H.H. Shugart and M.P. Stevenson, Climatic change and the broad-scale distribution of terrestrial ecosystem complexes. *Climatic Change* **7**, 29-43 (1985).
4. S.S. Batie and H.H. Shugart, The consequences of climate change on environmental resources: an economic and biological assessment. Prepared for workshop on Controlling and Adapting to Greenhouse Warming, Washington DC June (1988).
5. R.A. Park, M.S. Trehan, P.W. Mausel *et al. The Effects of Sea Level Rise on US Coastal Waters.* Holcomb Research Institute, Butler University, Indianapolis, Indiana, pp. 1-59 (1988).

Discussion Period 2

R. Pomerance (World Resources Institute)   Dr Shine, could I focus on HCFC 22? Ramanathan's prediction was that HCFC 22 might contribute around 6% of future warming between 1980 and 2030. You implied that he had underestimated the global warming potential of HCFC 22 by a factor of 2.5. In that case the CFC contribution rises to 15% over the next 50 years. Is that correct?

Dr Shine   Yes. Of course, Ramanathan *et al.* used the best data that were available. Since then there have been improved measurements which will bring HCFC 22 into the 0.1 K category, the same order of magnitude as nitrous oxide over that period.

Q   I would like to ask all the speakers about something that has worried me about their presentations. All the discussions so far have been that there is going to be an average change. Dr Oppenheimer spoke about the fact that there would probably be differential changes according to latitude. But what worries me are the extremes of climatic change. If the extremes occur more frequently, the potential effects will be greater than predictions based on the average. We should examine the frequency with which these extremes occur and their effect on particular ecosystems such as wetlands.

Professor Wigley   The point you make is relevant. If the temperature is changed by a small amount, the frequency of extreme temperatures will change by a much greater amount. The relationship between the two is non-linear, and it becomes even more non-linear if you consider back-to-back extremes. Michael Oppenheimer showed the frequency with which 100 °F would be exceeded in different US cities, and in many cases the number of days increased by a factor of 2 with a temperature change of only a few degrees Celsius. One sees a similar effect in calculating the probability of drought. According to one model, if carbon dioxide doubles, the probability of a severe drought in Kansas, which is currently one year in 50, becomes closer to 50 in 50.

Derek Cooper (Staffordshire Polytechnic) Carbon dioxide is produced by the burning of fossil fuels, yet fossil fuels derive originally from plants and therefore from atmospheric carbon dioxide. Does the geological record tell us anything about temperature before coal measures were laid down?

Professor Wigley There is an interesting anomaly in the global climate called the 'faint Sun paradox'. Three billion years ago, the Sun was only about 75% of its present luminosity, and yet we know that the Earth was not frozen over at that time. One explanation, not the only explanation however, is that all the carbon dioxide that is now locked up in limestones and fossil fuels was in the atmosphere. This carbon dioxide came from volcanic eruptions in the early days of the Earth's formation. These enormous quantities of carbon dioxide had a strong greenhouse effect that compensated for the fact that the Sun was weaker. The root of the problem, however, is what happens when a large chunk of the stored carbon is released into the atmosphere in the course of 75 or 100 years. It guarantees a very large change, very quickly.

Part 4
International Controls

10
The European Dimension

STANLEY CLINTON DAVIS
Commissioner to the European Communities

This is a timely conference on an important theme. Over the past 10 years the ozone layer has gone through the typical metamorphosis of an environmental problem. In the early 1970s the few voices expressing concern about the impact of CFCs on the upper atmosphere were regarded as being (by the polite) alarmist and (by the less polite) cranky. This was yet another example of bearded environmentalists building a thesis of global disaster on the basis of an obscure hostility to one of the most useful products of the modern age — in this case, the aerosol. In the more enlightened circles a few studies were set in train, but most of the world carried on with its spray cans as before.

Now, in 1988, the problem has come fully of age. We have a major, and epoch-making, international protocol on the subject, which is already widely seen as inadequate. We can be confident that by the end of the century the production of a key range of industrial substances will virtually have ceased. And we are waking up to the fact that the ozone layer problem is merely one aspect of major atmospheric changes brought about by man's activities which, unless we act swiftly and effectively, are going to have a devastating effect on the whole pattern of life on this planet.

Would it be too self-serving to draw the obvious moral from this history that mankind should listen harder to its environmentalists?

Anyway it is not my business here today to teach such general conclusions to so distinguished an audience — who I suspect largely share them anyway. I have been asked, rather, to explain the role that the European Community has played and is continuing to play in dealing with the ozone layer problem.

Ozone Depletion: Health and Environmental Consequences
Edited by R. Russell Jones and T. Wigley
© 1989 John Wiley & Sons Ltd

The Community is interested in the question from two aspects. One obvious aspect from which, by reason of its very nature, the European *economic* community has had to be involved, was the economic. The protection of the ozone layer requires binding international controls on the production, trade and consumption of CFCs and halons.

These are precisely the kind of matters on which the 12 Member States act together through the Community. It was especially important that they did so in this case. Europe produces and consumes about half of the world's CFCs. Thus, major European economic interests have been at play, and it has been most important that the European states act unitedly to ensure that the economic changes that are coming are spread fairly both within and beyond the Community's frontiers.

The second aspect is a much less obvious area of Community activity — the environment. Indeed, when the Community was founded, the environment was not even mentioned in the Treaties, an omission which was only formally rectified in 1987. But, in practice, it has been clear since virtually the beginning that environmental problems – with their strong transfrontier effects and heavy economic implications – are very much a matter for joint Community action.

Indeed, it was from the environmental point of view that the Community first addressed the ozone layer problem, in 1978. In that year a resolution was adopted which first emphasized the effect of CFCs on the environment. This was followed by concrete decisions in 1980 and 1982 to limit European CFC capacity, to cut the use of CFCs in aerosols, and to cut CFC emissions in a number of industrial sectors. As a result, European codes of good practice have been established for synthetic foams, refrigeration units and solvents, to reduce CFC emissions. We are now planning to update these and turn them into formal European standards.

However, of course, the ozone layer problem is not one that can be tackled at a European level. It is *par excellence* a *global* environmental problem, which requires global solutions. It is no use that this or that state, or even this or that regional organization should decide to cut CFC production if it is possible for that production simply to hop across a few frontiers and restart elsewhere. The only effective action is one that involves *all* the major producers and consumers.

At this point it is perhaps worth looking back to the world before Vienna and Montreal and reflecting on what the prospects were. We were dealing with the arguments of a body of opinion — the environmentalists — who, as I have already said, were regarded with polite derision in most establishment circles. We were dealing with a phenomenon that looked remote at best, the impact of which would not be felt for several years and about which there was extensive scientific disagreement.

We were dealing with a vast, and central, industrial activity whose significant reduction would cost companies and national economies many millions of dollars. And we were dealing with an act of international cooperation which had to involve everybody — communist and capitalist, producer and consumer, east and west.

Astonishingly, we did it; and I am proud to be able to say that the European Community played a central role. Regrettably we were not able to take a firm lead as the USA did. The USA has the advantages of a unified government approach and a high level of environmental concern nationwide. Europe does not.

What Europe did serve as, however, was what might be described as a test bed for the wider negotiation. The divisions among Member States — producers *versus* consumer, environmentally concerned *versus* environmentally unconcerned, industrially developed *versus* industrially developing — mirrored many of the divisions to be found in the wider negotiation. However, the European countries have the traditions of and institutions for mutual trust and cooperation which enabled these differences to be overcome within the Community. Once the Community had a common position, representing as it does the views and interests of the world's major trading bloc and preponderant producer and consumer of CFCs, that position was able to serve as a significant focus and stabilizer for the worldwide discussion.

Thus, while I cannot claim that the Community marched with the vanguard of the ozone layer relief army, I can claim that we provided much of the discipline and *esprit de corps* that maintained cohesion among the main body of the troops. And, moreover, when the Vienna Convention and Montreal Protocol had been negotiated we provided the framework of Europe-wide law that permitted them to be implemented swiftly, precisely and fairly throughout the Community.

I do not propose to go into the details of the Vienna Convention and Montreal Protocol. Other speakers will be doing that. I have already emphasized how implausible such agreements looked before we achieved them. Let me now move on to their implications for the future and for future Community action. I will take this at two levels — first on the particular problem of the ozone layer and then more widely as it is likely to affect our response to other global environmental problems.

On the ozone layer first, perhaps the right way to think of the Montreal Protocol is as a snowball that has rolled about halfway down the hill. The impulse towards worldwide cutbacks in CFCs and halons is now unstoppable and is growing rapidly. In the 14 months since the signature of the Protocol it has become abundantly clear and — what is more difficult — widely accepted in political circles that it is not enough. A recent European conference in The Hague concluded that cuts in CFC production of at least 85% before the end of the century are necessary to begin to reverse the damage that we have done. It was striking that when Community Environment Ministers debated the subject last week the disagreement was not over whether further cuts were required but over whether we should publicly announce a target of 85% now or should await more evidence to see if that is enough.

Perhaps as important is the impact of the new mood on the European chemical industry. Much of the resistance to Montreal came from large chemical companies and their sponsoring governments who were deeply worried about the effect of massive cutbacks in one of their major product areas. They now accept this as inevitable and are transferring resources and ingenuity out of defending the old into creating the new. You will all have seen press reports of progress in the development of CFC substitutes. Once these are available, I would expect to see industrial obstacles to a faster rundown of CFC production melt magically away.

Putting these factors together it is quite clear to me that by the end of the century worldwide production of the most damaging CFCs will virtually have ceased. The European Commission is already committed to achieving this. It is not clear how the renegotiation of Montreal will run or what precise targets it will set, but our destination is now fully visible. Within the Community we will be working for it both directly — through negotiation of further production and consumption cuts — and by a variety of indirect means, including looking for stronger controls to prevent CFC production moving outside the scope of the convention and the distinctive labelling of CFC-free products.

Let me conclude by looking at the wider impact of the ozone layer negotiation for international environmental policy. One way of putting this is to note that the Montreal Protocol at last makes it possible to achieve European Community style environmental agreements on a global scale. The Community, as I have said, has a long track record of binding legislation designed to tackle environmental problems that are felt Europe-wide. This legislation covers subjects ranging from car exhaust emissions to renewable packaging. In the light of Montreal it is now becoming realistic to think in terms of a range of worldwide environmental conventions to tackle the environmental problems that need to be dealt with on a global scale.

One such discussion, which is already well under way and which has been given a significant extra impetus by the ozone layer success, is the negotiation of a global convention on the transfrontier movement of hazardous waste. You all know the problem. The economic incentives simply to dump toxic waste in the developing world, rather than treating it properly in the developed world, are enormous, and the potential pollution of water tables, rivers and seas is terrifying. It is a practice that is growing rapidly and which can only be properly policed at a global level. In the peculiar way of the world, Montreal can be regarded as a significant step towards solving it.

However, the big test of the exemplary value of the Montreal Protocol will not be the waste problem, it will be the greenhouse effect. I note that you have already devoted an extensive session to this, so I will spare you any repetition of the details. We in the Commission see this as the big environmental challenge of the next few decades. We have just produced a major communication on the subject that initiates a phase of intensive work in the very wide range of policy areas — environment, energy, development, agriculture and so on — all of which will have to make their contribution.

It is, of course, the rising level of concern about the greenhouse effect — to which CFCs contribute some 20% — that has added to determination to cut CFC production right down. However, with the greenhouse effect, as with the ozone layer, Europe cannot tackle the problem by itself. On the other hand, without the European Community the problem would be significantly harder to tackle. The Community is now alert to the problem, has begun work on it, and will throw its full weight into finding the global solutions that are necessary.

Let me close on an optimistic note. Working on the environment, as I do, one is constantly dealing with problems and threats. One constantly sees warnings ignored and pressure groups successfully defending their narrow interests – often against the general good. At the widest level, one is constantly looking in vain for the sort of political courage and willingness to cooperate that is necessary if the race is to survive. Against that background the development of international environmental cooperation, first at a Community and then at a global level, shines out like a bright light in a dark world. Montreal is the clearest possible demonstration that mankind can cooperate to deal with the very real problems that he has brought to the planet. It is the foundation stone for a building that has to be erected rapidly and in difficult circumstances. At least the foundation is now there. We in the Community are working hard on the next steps of the construction.

11
The Montreal Protocol on Substances that Deplete the Ozone Layer: Its Development and Likely Impact

PETER USHER
United Nations Environment Programme

THE NEED FOR LEGISLATION

Chlorofluorocarbons are useful chemicals. Because of their benign nature, they lend themselves to a variety of domestic and industrial applications, which has ensured an increasing demand for their production since their first synthesis in the early 1930s. Current production estimates[1] are 12 000 metric tons of the CFCs and halons scheduled for control on the Montreal Protocol on substances that deplete the ozone layer.[2] The decision to phase out such seemingly valuable chemicals was not taken lightly, particularly when, in most cases, substitutes as effective and safe are not readily available. After all, CFCs are the most widely used refrigerants and foam-blowing agents. The non-toxic, non-inflammable, non-reactive properties of CFCs combined with easy liquefiability make them natural candidates for use as aerosol propellants. CFC 113 has had an increasing use as a solvent in the electronics industry, and the more recent discovery of halons has resulted in a meteoric rise in their production for use as a safe and effective fire extinguishant, which does not contribute secondary damage during the putting out of fires as do water and chemical foam products. Halon fire prevention devices are particularly in demand in computer rooms, record

---

Ozone Depletion: Health and Environmental Consequences
Edited by R. Russell Jones and T. Wigley
© 1989 John Wiley & Sons Ltd

archives and for use by the military.[3] There is every reason to believe that, environmental effects apart, the production and use of CFCs and halons would have continued to rise exponentially in the foreseeable future, particularly as the development aspirations of the non-industrialized countries began to be realized (Figure 1).[3]

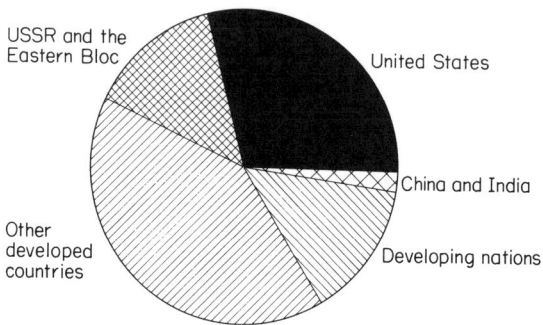

Figure 1   A global problem. (Source: EPA[3].)

The demise of the CFC growth industry was brought about by classical scientific process – elegant theoretical prediction enforced by rigorous monitoring, scientific experiment and eventually confirmation beyond reasonable doubt by atmospheric observation. Yet science has not always been successful in persuading commercial interest to replace profit by environmental altruism. Why so in this case, particularly when the problem is perceived as global in impact yet regional in origin? The greater part of CFC production and use is confined to the industrialized world, although the potential for future use would naturally spread and increase as more countries develop and become more affluent. The reason is probably that an international mechanism with the mandate of, and capacity for, addressing global environmental issues existed. The realization of some ideals of the 1972 Stockholm Conference on the Environment in a relatively short period was one of the more satisfying by-products of the ozone layer protection programme.

The ozone destruction theory was initially proposed in 1974.[4] It suggested that the relatively stable CFC molecule would survive unchanged in its emission and transport through the troposphere, with its non-reactiveness and insolubility ensuring that the greater proportion of such releases would eventually reach the stratosphere, wherein lies the major part of the ozone layer. Molina's and Rowland's proposal was in essence that reactive chlorine (and nitrogen from natural and manufactured chemical sources) could be released by photochemical action on CFCs, and that by a relatively simple chemical reaction natural ozone could be reduced to molecular oxygen with the reactive chlorine atoms remaining unchanged to catalytically render tens of thousands of additional ozone molecules subject to destruction.

The ozone layer is the earth's shield from ultraviolet radiation of wavelengths up to approximately 320 nm. UVC (200–290 nm) is lethal to living things and is almost totally absorbed by ozone. UVB (290–320 nm) is mostly absorbed.[5] However, the small portion that does reach the Earth is environmentally damaging. It affects human health, causing skin cancers, inducing cataracts and damaging the body's immune system. It affects plants and animals, reducing plant yields (including those of important food crops) and it is particularly severe on juvenile forms of marine life. UVB is also a factor in acidic pollution and it can degrade plastics and other materials used in construction.[6] Even small reductions in ozone conentrations can have significant effects. It is calculated that a 1% decrease in total column ozone would increase the amount of UVB reaching the Earth by 2%. This radiation amplification effect would be supplemented by a biological amplification effect, which could increase certain types of skin cancer by about 2.5–4.5%. Predicted ozone depletion[6] would exceed 10% before the middle of the next century if CFC emissions were to grow at 2.5% per year over this period. Production growth since 1983 has been 5–8% per annum (Figure 2).

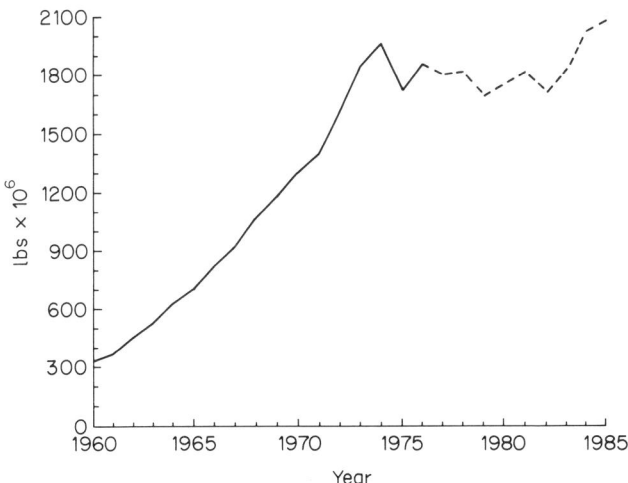

Figure 2 Approximate total world production of F 11/12/113. (Source: EI du Pont de Nemours & Co.)

There is another important effect of ozone layer modification. Stratospheric ozone loss will, to some extent, be compensated by ozone increases in the troposphere as more solar radiation penetrates into the lower atmosphere

(Figure 3).[7] Ozone is an important greenhouse gas and thus will enhance the expected climate change forecast to result from increasing carbon dioxide, nitrous oxide, methane and, of course, chlorofluorocarbons in the atmosphere. Even if total ozone can be stabilized by implementation of the Montreal Protocol or through even stronger regulatory measures, there will be a guaranteed contribution to global greenhouse warming lasting for decades and, perhaps, for centuries. Only the ozone depleting potentials of substances considered for regulation under the Protocol were taken into account during the negotiations,

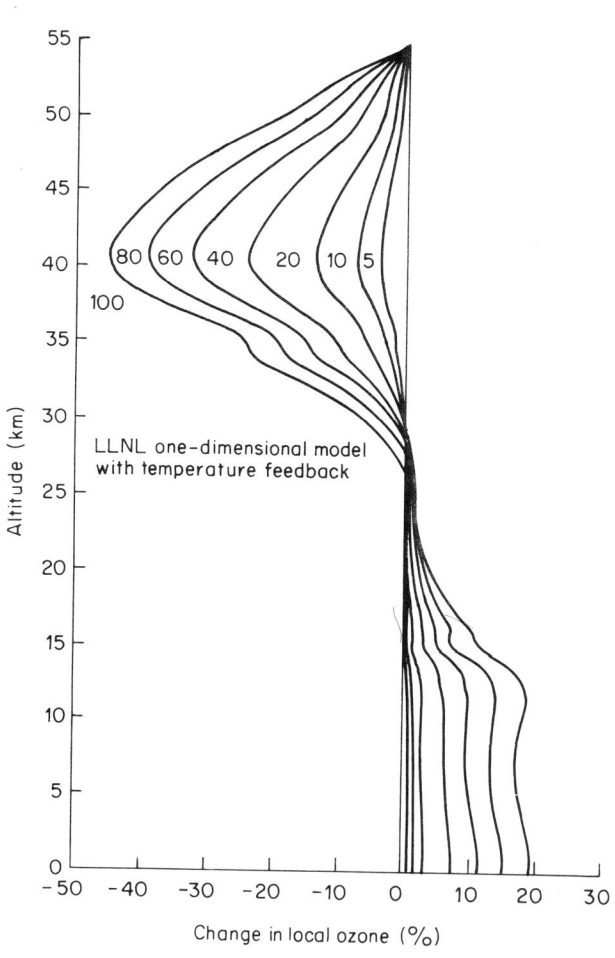

Figure 3 Calculated percentage change in local ozone at selected times (5–100 years) assuming CFC emissions begin at 1980 rates and increase at 1.5% per year, methane increases at 1% per year, nitrous oxide increases at 0.25% per year, and carbon dioxide increases according to the DOE scenario. (Source: NASA.[7])

not their greenhouse warming potentials. 'Acceptable' substitutes for the regulated substances need to be assessed for their total environmental effect, including their potential contribution to the state of the climate, and not just for their 'ozone-friendly' character. Nor will it be fair to industry if environmentalists offer a seal of approval to a substitute for regulated substance only to withdraw it in the context of similar legislation that may be developed to protect the Earth's climate.

ASSESSMENT OF OZONE LAYER DEPLETION

In 1976, UNEP was still in its first decade of operation. One of the operational elements was its outer limits subprogramme. Outer limits is the concept that there is a limit to which any part of the natural environment can be altered by human intervention, beyond which irreversible and unacceptable change occurs. The risk to the ozone layer was deemed such that by its depletion an outer limit would be transgressed. In 1976, concerned by the Molina–Rowland hypothesis, the Governing Council decided that UNEP should arrange for international consideration of the issue. In 1977 UNEP, with assistance from the US Government, convened an International Conference of Experts on the Ozone Layer.[8] The outcome was a general expression of concern and a requirement to study the issue further. The principal product of the Conference was a World Plan of Action on the Ozone Layer. Its theme was encouragement of research into atmospheric processes that control the concentration of ozone in the stratosphere and of the modification of the ozone layer by human intervention through emission of precursor trace gases, particularly the halogen-containing compounds indicted in the ozone destruction process. The plan also provided for the assessment of the effects of ozone layer change.

The specialized agencies of the United Nations system were designated to arrange for the component parts of the assessment of ozone layer depletion and its impact. The World Meteorological Organization (WMO), for example, would have responsibility for the monitoring of atmospheric constituents and research into atmospheric processes; the United Nations Food and Agriculture Organization (FAO) was to arrange for the better understanding of UVB incidence on plants and ecosystems; the World Health Organization (WHO) was to advise on the risk to human health; and the International Civil Aviation Organization (ICAO) on the effect of air transport, particularly of supersonic aircraft operating in the lower stratosphere, which in the mid-1970s was considered an area of probable major development by aviation industry. UNEP's role was to be an overall coordinating one and, to assist it in this task, it was recommended that a special Coordinating Committee on the Ozone Layer (CCOL) should be established. This UNEP body would have a membership comprising those organizations implementing the World Plan of Action on the Ozone Layer and representatives of countries having major research programmes concerned with the ozone layer, its modification and the effects of such changes.

In the event, with the particular exception of the WMO, most of those organizations designated to lead the assessment process demanded by the Plan declined to participate or to establish specific programmes necessary for the implementation of the Plan. UNEP was informed that either low priority was given to the issues within Agency programmes or that the Agency discounted the risk as insignificant. However, some non-governmental organizations, notably the Chemical Manufacturers Association (CMA) and the International Council of Scientific Unions (ICSU), indicated their support of the developing UNEP programme and were invited to join the selected Governments at the first session of the CCOL in Paris in 1978.

The CCOL has been convened eight times. Its composition has not changed apart from the inclusion of a small number of Governments which expressed interest in becoming members. The more or less annual product was an assessment of ozone layer modification and its impact, published by UNEP in its ozone layer bulletins.[9] The assessments were based on the most recent research results obtained by members, and provided a regular evaluation of the trend in the atmospheric abundance of ozone and other trace gases influencing its concentration in the atmosphere — methane, nitrogen oxides, carbon dioxide and monoxide, and the chlorofluorocarbons. CMA supplied data on the production and use of CFC 11 and CFC 12 by its members and, when possible, estimates of the production of CFC by non-CMA members. The CCOL also provided forecasts of ozone layer depletions based on emission trends of the trace gases suspected of modifying ozone. In the early years, these estimates varied widely, upward of 15% one year and lower than 5% the next. This, unfortunately, tended to shake confidence in the scientific community's ability to provide reliable information on the ozone layer and, in some countries, the threat of ozone destruction was dismissed as irrelevant. The sharp downward trend in CFC production which occurred as a consequence of the ozone destruction hypothesis and action by some countries, notably the United States of America, to limit the non-essential use of CFCs — for example, as propellants for cosmetic products packaged in aerosol spray-cans — was reversed. With more uses being found for CFCs and increasing use of the gases as foam-blowing agents, CFC production in 1986 matched the peak production of ten years earlier and annual production growth of CFC 11 and CFC 12 reached 8–11%. Production of other CFCs was also rapidly rising.

The unreliability of scientific prediction was, in fact, a consequence of the growing awareness of the complexity of the issue. More and more chemical reactions having a bearing on ozone concentrations were discovered. The non-linearity of ozone destruction mechanisms depending on the relative concentrations of active chemical constituents of the stratosphere, the highly coupled nature of the chemical reactions, the growth of ozone-increasing chemicals in the atmosphere as well as ozone-destroying ones, re-estimates of chemical reaction rates and the influence of the solar cycle were only a few of the additional complications to be mathematically modelled by the analysts. The ozone-destruction theory of Molina and Rowland, however, remained intact. Ozone is destroyed by active chlorine released by the decomposition of chlorofluorocarbons. By 1981, based on CCOL estimations, Governments, albeit some reluctantly, decided to work towards the development of a global agreement for the protection of the ozone layer. UNEP was charged with arranging for this by its Governing Council[10] and UNEP's Environmental Assessment Service which maintained a continuing programme of scientific assessment through the UNEP CCOL.

THE VIENNA CONVENTION FOR THE PROTECTION OF THE OZONE LAYER

By establishing an *ad hoc* working group of legal and technical experts nominated by concerned Governments, UNEP initiated the process that was to lead to the adoption in Vienna in March 1985 of a global framework convention for the protection of the ozone layer. The use of the term 'framework' recognized a future need to develop specific regulatory protocols that would be binding on Parties to the agreements.

The Vienna Convention was a precedent-setting agreement. Although other internationally agreed environmental legislation had preceded it, never before had there been acknowledgement of a global problem of the future rather than the present. Before 1985, ozone depletion, although predicted to have occurred, could not be verified by observation. Instrumental insensitivity and large natural variability in ozone concentrations prevented confirmation of the threat. Efforts were concentrated on detecting stratospheric depletion rather than total ozone change as it was likely that initial signals of ozone destruction would show in the region most vulnerable to chemical decomposition of ozone – about 20–40 km above the Earth's surface. As it happened, the greatest change, so large as to be almost unbelievable, was observed over Antarctica.[11] The unpredicted Antarctic Ozone Hole (Figure 4) was recognized at the same time that the final stages of the agreement of the terms of the Vienna Convention.[12] It had no impact on the content of the Convention, but was to become a significant motivator in shaping the future protocol to control the emission of substances suspected of ozone layer depletion, even though confirmation of the role of CFCs in the destruction of Antarctic ozone was not obtained until after adoption of the Montreal Protocol.

The Vienna Convention obliges Parties to the agreement to take appropriate measures to protect human health and the environment resulting from the effects of human destruction of the ozone layer. The main theme of the agreement is to encourage intensified action towards a better understanding of the atmosphere and its processes through research and observation; cooperation among states on

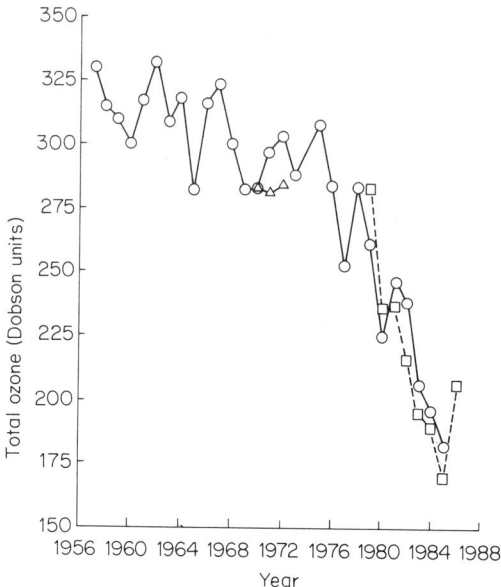

Figure 4 October monthly mean total ozone measured over Halley Bay Station, Antarctica. ○, Dobson; □, SBUV; △, BUV. (Source: NASA.)

scientific, legal and technical matters; and free exchange of information concerning scientific, technical, socioeconomic, commercial and legal aspects of ozone layer protection. Much of what was recommended, at least as far as scientific matters were concerned, was already in operation under the CCOL, but global emission control beyond that already undertaken on a national basis or within international economic groupings would require a formal legal undertaking by States to cooperate in activities and pool information necessary to determining and applying the appropriate actions to protect the ozone layer.

The European Economic Community and 22 countries signed the Vienna Convention for the Protection of the Ozone Layer on 22 March 1985 but, notwithstanding the slow process of amending national laws to take account of international agreement, there was little rush to formally ratify the treaty and thereby oblige compliance with its conditions, innocuous as they generally are. States adopted a 'wait and see' policy, holding back until the future Protocol on regulatory measures was developed. Attempts to agree on a protocol for adoption at the same time as the Convention failed completely and UNEP was asked by its Governing Council to reconstitute the Vienna Group of legal and technical experts to elaborate a protocol on chlorofluorocarbons to the Vienna Convention.[13]

DEVELOPMENT OF A PROTOCOL ON SUBSTANCES THAT DEPLETE THE OZONE LAYER

The protracted negotiation of the Vienna Convention confirmed the sensitivity of some nations with regard to what was considered necessary to protect the ozone layer. Opinion varied from a simple freeze on CFC 11 and CFC 12 to calls for a total ban on the full range of ozone-destroying substances. Considerable preliminary work would be necessary to reconcile differing national opinions which, in many cases, appeared to reflect commercial interest rather than environmental concern.

First steps included the convening of a two-part workshop to consider what elements a protocol to control CFC emissions might contain. The first workshop, in Rome,[14] examined national production, use and trade statistics for CFCs. It also examined national and regional laws and agreements already in place or scheduled for implementation. Although this meeting considered little more than statistical data and current legislative procedure, the meeting was distinguished by disagreement and bad temper, auguring poorly for future negotiation. The second workshop at Leesburg, USA[14] was, however, exemplary of a successful meeting. Although faced with the theoretically more difficult task of comparing different models for regulatory measures and assessing their effectiveness, efficiency and fairness, participants came away from the meeting embued in the 'spirit of Leesburg', an evocative term coined by the Workshop Chairman, Ambassador Richard Benedict of the USA, and summarizing a remarkable degree of unanimity on the need for an effective treaty and a willingness to give sympathetic consideration to the four different control formulae submitted to the meeting by Canada, the EEC, USSR and Sweden.

The Vienna Group was brought together quickly by UNEP in three sessions[15] separated by the shortest possible intervals, allowing time only for preparation of agendas and other conference documents. The CCOL met in two sessions[16] to summarize the latest scientific understanding of the issues published in a three-volume document by NASA,[17] reflecting the opinion of several hundred international scientists. This information was made available to the Vienna Group. Scientific opinion was firm and uncompromising on the issue. The atmosphere was being subjected to a vast, uncontrolled experiment and, unless emissions of CFCs and other potential ozone-destroying substances were curtailed, ozone layer depletion of a magnitude that could induce an unacceptably high environmental risk, including damage to human health, would result.

Evidence of further losses of Antarctic ozone was also confirmed by satellite and ground-based observations which, although at the time could not be attributed to human intervention, demanded a precautionary approach.[11,18] Scientific opinion was divided on what caused the seasonal ozone hole in the Antarctic stratosphere and it was variously attributed to variability in the solar cycle, the unique meteorology of the polar region characterized by a cyclonic vortex isolated from the general global circulation, and chemical decomposition. A major scientific campaign involving ground, aircraft, balloon and satellite measurements, was to be mounted in the Antarctic spring (October–December) of 1987, with the results to be compiled and published in early to mid-1988; unfortunately, out of step with the protocol process, which was seeking conclusion in September 1987, given the offer of the Government of Canada to host a Conference of Plenipotentiaries to adopt a Protocol at that time.

The Vienna Group sessions in Geneva, Vienna and again in Geneva did not fully respond to the optimism of Leesburg. Deep divisions along national lines were apparent and many were still unresolved as the Montreal Conference approached. The major areas of disagreement were whether regulation should be applied to production or production capacity, emissions or consumption of CFC (an expression to reflect trade in CFCs in addition to production and use). A second stumbling block was whether or not CFC production should be frozen, cut back or, as some would preferably have it, eliminated entirely. A third issue was what chemicals should be regulated; CFC 11 and 12 only or a range of ozone-destroying chemicals including the halons (bromine-containing compounds with calculated ozone destroying capacity of 3–12 times that of the reference fluorocarbon CFC 11). The less developed countries, initially uninterested in the ozone layer question, attended Vienna Group meetings in increasing strength. Such countries were keenly aware that some of the control strategies, if adopted, could limit their access to chemicals important to their development aspirations. There was an increasing need for refrigeration and air conditioning, and in the newly industrialized countries there was in some cases even a growing requirement to use, make and export products made with, or containing, CFCs. Per capita use in the developing world is approximately a tenth or less of that used in Western society, which has a per capita consumption of 1–2 kg/year. Developing countries demanded a fair deal from the negotiations – access to the suspect but useful chemicals until such time as substitutes were available. Another thorny issue was trade, both among prospective parties to the treaty and between parties and non-parties. What trade barriers should be erected? How should they be enforced and what penalties should accompany the contravention of trade rules?

The first three issues could be adjudged to be technical ones. After all, the purpose of having a treaty was to protect the ozone layer from unacceptable modification. Anything less would be a waste of time. There is also validity in the argument that cut-backs of CFCs beyond those necessary to stabilize the ozone layer within an acceptable depletion limit are economically debilitating. The usefulness of CFCs is incontrovertible and substitutes can be dangerous (inflammable and toxic) and expensive. One major problem with ozone-related research is the degree of uncertainty associated with prediction. Theoretical model prediction is often contradicted by observation. A preoccupation with scientific integrity and an over-qualification of research conclusions easily allow misrepresentation of the intentions of the scientists concerned. The Vienna Group negotiations provided a classic example of virtue exploited. It was argued that such was the disparity of depletion predictions from the various mathematical models that such calculations could be discounted, and that until more reliable results could be obtained only limited precautionary CFC control should be agreed — freeze now and wait for scientific confirmation. This argument is wrong on two counts. Continue to pollute until such pollution is proved to be harmful not only defies the principles of natural justice, but also contradicts the argument fundamental to the whole issue. In the case of the ozone layer, confirmation of the ozone depletion theory would be coincident with being no longer able to repair the damage. The quantity of pollutant in the atmosphere and its long residence times would ensure our helplessness to respond. Secondly, the disparate results were likely to result from different assumptions of trace gas and economic futures rather than fundamental differences in the models themselves. Given the probable similarity of the few models applied to the ozone layer problem, putting the same data in would mean getting the same results out.

This was confirmed in a hastily arranged meeting of modellers by UNEP in Wurzburg. Each model team was given an identical set of base data on trace gas emission rates and projections of demand for CFCs by both the industrialized and developing world. The meeting not only confirmed model consistency, it also made sobering predictions of future atmospheric change.[19]

For all the given, not unrealistic scenarios, ozone layer depletion would continue and accelerate unless the full range of ozone-destroying substances was regulated. Also, a freeze in emissions was insufficient to stabilize the atmosphere unless it was truly worldwide (effectively, significant cutbacks in the developed world to compensate for growing demand in the developing world).

The third meeting of the Vienna Group in Geneva confirmed through its technical experts the findings of the Wurzburg meeting, and subsequent negotiations accepted the 'CFC basket' and 'reduction formula' approach as a basis for regulatory measures. The spectre of the Antarctic ozone hole also haunted the negotiations. Irrespective of whether or not it was man-made, it had not been foreseen. The atmosphere was not fully understood and the risk of that giant uncontrolled experiment was plain to see.

Final negotiations by technical working groups began on 6 September 1987 in Montreal, and on 15 September a 2-day Conference of Plenipotentiaries (persons with assigned legal authority to act on behalf of their governments) saw the adoption and signing by 23 countries of the Montreal Protocol on Substances that Deplete the Ozone Layer.

Such was the sensitive nature of the negotiations, which culminated in agreement on the Protocol literally hours before its consideration at the conference of Plenipotentiaries, that the final text contains several ambiguities and reflects issues still to be resolved. No editorial process was permitted on the document for fear that the carefully negotiated compromises might be altered in meaning if expressed differently, even if more grammatically. Nevertheless, the agreement is not a fragile one. Its requirements are for powerful regulatory measures, which yet take account of particular national and regional grouping needs. In its compromises, it has still remained true to the requirement of environmental integrity and, while satisfying no-one completely (being accused by environmental groups as too weak and industrial lobbies as too strong), it is accepted as being able, when implemented, to provide good protection for the ozone layer. Figure 5 shows calculated ozone change under the protocol compared with that of continued CFC growth. The Protocol has the virtue of flexibility, allowing it to be adapted rapidly should new scientific information indicate a need for change; tightening or relaxing regulations as circumstances dictate. It is remarkably fair to developing countries, providing them with access to the controlled substances for a long grace period, even as significant cutbacks operate in the industrialized world. It provides trade and chemical availability benefits, such that there is an incentive for non-parties to ratify the treaty at an early stage.

The following is an Article-by-Article examination of this precedent-setting environmental agreement, pointing out the various strengths and weaknesses of the Protocol and indicating the Programme for its implementation.

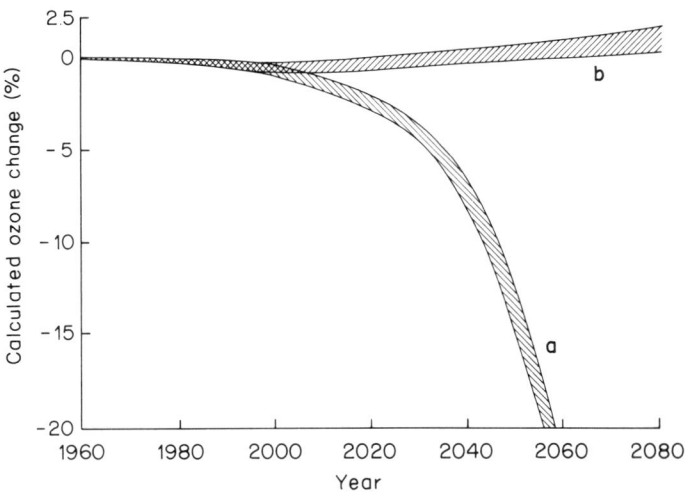

Figure 5 Comparison of (a) growth *versus* (b) Montreal Protocol. Growth in (a) is assumed to be at 2.5% per year compounded growth. (Source: EI du Pont de Nemours & Co.)

THE MONTREAL PROTOCOL

Preamble

This provides the rationale for elaborating a protocol on substances that deplete the ozone layer. Less cautious than the similar preamble to the Vienna Convention, it recognizes that emissions of certain substances can significantly deplete and otherwise modify the ozone layer in a manner that is likely to result in adverse effects on human health and the environment. The Convention, on the other hand, stresses the need for scientific knowledge and research as a prerequisite for action. It makes no reference to substances that might deplete the ozone layer, but only oblique references to activities that might modify it. Definitions are provided for 'Convention', 'Parties', 'Secretariat', 'Controlled substances', 'Production', 'Consumption, 'Calculated levels of production', and 'Industrial Rationalization'. Of note are the following:

(1) *Controlled substances* are those listed in an Annex to the protocol, and specifically prohibit the exemption of substances, even when contained in a mixture with non-controlled substances, and where the overall ozone depleting potential (ODP) of the mixture is small. An example of this is the aziotropic substance CFC 502, a mixture of CFC 115 (a controlled substance) and CFC 22, which is not included in the annex.

(2) *Production* is the amount of controlled substances produced, minus the amount destroyed by technologies to be approved by the parties. The credit for destroyed substances is currently little above zero. Parties have yet to approve technologies for destruction, of which burning is one of the few successful methods employed. There is an incentive here for producing nations to develop and apply CFC destruction technologies to augment allowable production.

(3) *Consumption* is defined as production, plus imports, minus exports of controlled substances. The atmosphere recognizes emissions of controlled substances as it is emissions that will regulate the state of the natural ozone layer. The negotiators agreed that 'consumption', as defined, would provide the best quantifiable estimate of emissions.

(4) *Calculated levels of production*, imports, exports and consumption refers to calculations of these elements using the formula agreed in Article 3.

(5) *Industrial rationalization* It was recognized that the economic efficiency of producing plants might be compromised as consumption limits began to bite. The transfer of production quotas from one country to another was seen as a way of offsetting this problem and, at the same time, ensuring an adequate supply for developing country non-producers without significant market price increases – the complementary response to a shortfall in supply.

Article 2: Control measures

This article is the *raison d'être* of the Protocol and distinguishes it from its parent Vienna Convention in its obligation on Parties to reduce consumption of substances that deplete the ozone layer.

Paragraph 1 provides a timetable for a freeze on consumption and production at, or below, each Party's consumption in 1986. Assuming an entry into force of the Protocol at its intended date of 1 January 1989, production and consumption for the 12-month period beginning 1 July 1989 must not exceed calculated or estimated production and consumption for 1986. A rider to this restriction allows production and consumption to be increased by 10% of 1986 levels, but only to satisfy the basic domestic needs of low-consuming countries and for the purpose of industrial rationalization between countries. A question arises as to what is meant by 'basic domestic needs', which is not defined in Article 1. It is also queried whether the right to export is a basic domestic need. Certainly some countries, significantly higher consumers than producers based on a trade in goods manufactured with or containing CFCs, would not be affected as such products lie outside the definition of controlled substances. However, a trade in raw chemicals fed by the allowable production increase above freeze levels lies outside the spirit of the Montreal Agreement, even if debatably within its letter.

Paragraph 2 outlines the control schedule agreed for group 2 substances. Again, assuming 1 January 1989 as the entry into force date, then by 1 February 1993 a second scheduled list of substances to be controlled, the bromine-containing halons, have their consumption and production restricted at or below the 1986 levels. The same permitted increase in production is allowed – 10% for satisfying basic domestic needs of developing nations or industrial rationalization.

Present growth in halon production is estimated at 5–10% per year. Thus, a 4-year grace period may eventually mean a need to cut production by 20–40% to satisfy the regulatory measures.

Paragraph 3 outlines the phase-down schedule for the controlled CFCs. By 30 June 1994, Parties will have had to reduce their production and consumption by 20% of 1986 levels. A 10% cost production increase above 1986 levels may be added to the new calculated allowable production level for basic domestic needs and industrial rationalization purposes.

Paragraph 4 outlines the further reduction phase, a semi-automatic step contrasting with the inescapable nature of the earlier freeze and first-phase reduction. The new alllowable consumption is 50% of the 1986 reference year. However, the allowable rise in production is 15%. This increase reflects an anticipated greater need for industrial rationalization under low permitted production limits. The semi-automatic condition upon this regulation is that it will come into effect unless two thirds of the Parties representing two thirds of total global production decide otherwise. The basis for such objection would be made on assessments of the state of current scientific knowledge arranged for under Article 6.

Paragraph 5 considers the special situation of low-consuming countries, which might require an industrial rationalizing programme to ensure supply of the controlled substances under conditions of forced closure of existing small producing plants because of their being no longer economically viable. Such countries, with a calculated level of production of less than 25 kt, could transfer to, or receive production credits in excess of that proscribed in the earlier paragraphs provided that the combined production of the Parties concerned does not exceed the limits specified earlier. For example, an arrangement between Canada and the USA might protect the consumption needs of the former, which provides only 2.5% of global CFC production.

Paragraph 6 acknowledges the existence of countries with centrally planned economies and subject to legally binding obligation for production and supply of the controlled substances to their trading partners. It also allows production facilities, arranged for before 26 September 1987 but not yet operating, to include its projected production figures in its 1986 calculated level of production. Additional qualifications prevent new production from raising that country's per capita consumption above 0.5 kg per capita, and insist that the new facilities must be on stream before 1991. This clause was included to satisfy the specific needs of the USSR.

Paragraph 7 ensures that legitimate production transfers for the purpose of industrial rationalization are notified to the Secretariat for accounting purposes.

Paragraph 8 satisfied the particular circumstances of the European Community acting as a regional economic integration organization with operating agreements on trade among the members not accounted for by the regulatory measures specified in early paragraphs of this Article. The rules allow member states jointly to fulfil their obligations in respect of consumption, limiting the total combined figure for the States to the aggregate of allowable consumption limits for individual members. The agreement requires that all member states and the parent organization are Party to the Agreement.

Paragraph 9 provides the mechanism whereby ODPs may be re-evaluated and/or further adjustments of control levels can be made. Agreement on such changes would be as far as possible by concensus. A voting procedure would be used as a final resort in the case of no concensus. Voting restrictions are somewhat less stringent than those required to upset the semi-automatic second reduction control phase to 50% of 1986 reference levels; two thirds of the Parties, representing 50%, rather than 75%, of total consumption. Nevertheless, the decisions will become binding on all Parties within a prescribed time of the notice being communicated by the Depositary.

Paragraph 10 provides procedures for amending the list of controlled substances by adding to, or subtracting from, it and the regulations that should apply to such substances.

Paragraphs 2, 3, 4 and 6 specify controls for Parties other than those operating under Article 5. The Article 5 parties are the developing low-consuming countries, which choose to take advantage of special provisions listed in Article 5. These will be discussed later.

The question arises as to what is the quantitative estimation of global consumption for the reference year. For practical purposes, global consumption equals global production of the 172 countries, plus 11 British Dependent Territories and the European Economic Community. The 172 countries referred to above are divided into 40 developed and 132 developing countries based on several reference lists (no definitive list of developing countries actually exists). Sixty countries reported 1986 production data. UNEP made estimates of data for 28 countries.[20] From this it was deduced that the 1986 reference figure for global consumption is 1 197 000 tonnes. The allowable increase in production for industrial rationalization and stock availability to developing countries is therefore about 120 kt/year. The reduction schedule for CFCs should approximate to those in Table 1.

Table 1

Control measure	Reference amount (kt)	Allowable global consumption (kt)
Freeze	1200 + 120 (10%)	1320
20% reduction	960 + 120 (10%)	1080
50% reduction	600 + 180 (15%)	780

Reference level consumption figures for Group 2 substances are not yet calculated.

Article 3: Calculation of control levels

The article specifies the formula for calculating production (P), which requires the annual production of each controlled substance to be multiplied by its ozone depleting potential. The products are then added together to obtain the overall production of the controlled substances. The calculation of imports (I) and exports (E) is made in the same way as for calculated levels of production using ODPs as factors to adjust the individual quantities of substances produced.

Consumption (C) is then calculated using the formula:

$$C = P + I - E$$

except that from 1 January 1993 no credit for exports to non-parties will be given.

The formula provides considerable flexibility of choice to Parties in what substances they produce or use — either smaller quantities of significant ozone destroyers or larger amounts of less-destructive substances on the schedule. The formula ensures that the impact on the ozone layer is unchanged regardless of the commercial decision taken within the context of Protocol compliance.

Article 4: Control of trade with non-parties

Parties are allowed a grace period of 1 year to discontinue imports from non-parties to the Protocol. This does not, however, prevent the importation of products containing the controlled substances, but only the import of bulk chemicals. A list of such products is to be drawn up and annexed to the Protocol within 3 years of entry into force. Such products from non-parties will be banned within a further year. The provision is, however, non-binding in that individual parties may object to compliance with the list in accordance with agreed procedures for doing so. This weakness in the Protocol reflected disagreement during negotiations and reflects the present impossibility of fully assessing the implications of a mandatory ban on products still to be determined. A specific example of this dilemma might be the risk of cutting off the supply of an 'essential' product presently available from a state that remains outside the agreement when no appropriate alternative is obtainable from a legitimate source.

In practice, it is thought that states would be restrained in their objections needing a strong argument to justify non-compliance with the 'spirit' of the Protocol.

Exports to non-parties is also banned from 1 January 1993. Such stringent restrictions in trade provide a strong incentive for states to become Party to the Protocol or risk the loss of their export markets or the cutting off of their supply of chemicals and products to be controlled under the treaty. Even more rigorous trade restrictions are contemplated — a possible downstream control of the movement of products made with, but not containing, the controlled substances between States-Parties and non-parties to the treaty.

Paragraphs 5 and 6 restrict the transfer of production and use technology to non-parties apart from those that improve the containment, recovery, recycling or destruction of the controlled substances and promote the development of alternative substances. Subsidies and credits, etc., that would facilitate the production of controlled substances in a similar manner is not to be provided to non-parties. However, since in practical terms such measures would be unenforceable, the provisions are expressed in non-binding language. A country complying with the regulatory measures but not formally a Party, could, with the agreement of the Parties, be exempt from import restrictions. This ensures conformity with the General Agreement on Tariffs and Trade (GATT).

Article 5: Special situation of developing countries

Model projections of ozone layer sensitivity to future emissions of ozone destroying substances (Wurzburg) confirmed the sensitivity of the atmospheric response to growth projections for developing countries. It was concluded that a treaty without global compliance would fall short of environmental needs. Equally, it was appreciated that the developing countries had legitimate requirements connected with their development aspirations, which would elevate their relatively low consumption of ozone-layer-depleting substances. Compensatory cutbacks by the industrialized world would probably be insufficient even if draconian in character. Developing countries would need to play their part, and reduction formula, treating such states preferentially, would have to be devised if such countries were to be included within the confines of a legislative agreement. A large number of developing countries took part in the final stages of the negotiation process and contributed to devising the fair and equitable solution expressed in the Article.

The special provisions agreed allow a 10-year delay in compliance to meet basic domestic needs provided the developing country consumes less than 0.3 kg per capita during that period. Low-consuming countries not recognized as developing would not be entitled to this advantage although they could operate under the special provisions of Article 2 for countries producing less than 25 kt of Annex I substances to facilitate industrial rationalization.

The reference level for 'consumption' for developing countries is either 0.3 kg per capita or the average of each country's consumption for the period 1995–97, whichever is the lower.

A problem arises in that this calculated reference level is for all controlled substances, i.e. those in Groups 1 and Group 2 combined. The control schedule that is now to apply treats these groups differently. Therefore, it will be necessary to disaggregate the overall consumption reported by developing countries into Group 1 reference levels and Group 2 reference levels in order to specify when and how the deferred control schedule is applied.

In practice, this may present less of a problem than at first sight. The anticipated phase out of controlled substances and the increasing availability of substitutes could mean that most developing countries will not be using CFCs by the date on which it becomes necessary to begin the phase-out schedule, which for Article 5 countries require effecting a freeze in consumption of Group 1 substances by 1 July 2000. Theoretically, developed countries exceeding the 50% of 1986 production for satisfying basic domestic needs of developing countries should, in 1999, drop consumption to 50%.

The developing countries' access to alternative substances and technology from other Parties is guaranteed, and the provision of subsidies, aid, credits, guarantees or insurance programmes denied to non-Parties provides an attractive inducement to ratification.

Theoretically, the developing countries still hold the fate of the ozone layer in their hands. Should the 132 developing countries increase their per capita consumption to the maximum 0.3 kg, the chlorine burden on the atmosphere could double. However, for this to be possible new large-scale producers/exporters would have to appear, as under the reduction schedule for current producers only 120–180 kt will be available for both industrial rationalization and transfer to developing countries to meet basic domestic needs. It is thought that large-scale investment in products to be phased out or superseded by alternatives would be economically unjustifiable, particularly as by 1997 such countries would, as Parties, be obliged to consumption and production phase-down. Nevertheless, China, India and perhaps a few other developing nations have high enough populations and industrial potential to strain the expectations from the Protocol as an environment safety net, particularly if such countries were to stay outside the Protocol, despite the trade restriction and loss of access to CFC production, and use technology that would automatically follow.

Article 6: Assessment and review of control measures

The efficacy of the control measures will be reviewed by the Parties in 1990 and at 4-yearly intervals thereafter, guided by expert assessments of scientific, environmental, economic and technical aspects. To ensure that the Parties are sufficiently informed to make the necessary 1990 review, UNEP is currently arranging for the four assessments to be completed in 1989.

Article 7: Reporting of data

This article provides the mechanism whereby parties can provide statistical data or, in their absence, estimates of consumption and production of, and trade in, the controlled substances. However, for the Protocol to enter into force in accordance with Article 16, which requires ratification, or its equivalent, by 11 states accounting for at least two thirds of 1986 estimated global consumption, the statistical or estimated data is needed now rather than within 3 months of individual states becoming a Party to ensure a legal coming into force of the Protocol. UNEP as the interim secretariat has therefore collected the production data and made estimations in cases of non-reporting or data absence. It is now accepted that entry into force will be effected when 11 or more states with a combined production of 800 kt of the controlled substances complete their ratification process. To the present, eight countries representing 32% of global production have ratified. These are Canada, Egypt, Mexico, New Zealand (excepting Cook Island and Nive), Norway, Sweden, Uganda and the USA. Two others, Japan and the Ukranian SSR, have respectively approved and accepted the Protocol. The ten countries that have deposited their instruments of ratification represent 48% of global consumption. Apart from a minimum of one more country ratifying the agreement, a further 19% of global consumption should be represented by new Parties. This can be achieved by the European

Community becoming Party *en bloc* or even by three of the larger producers among its member states, should individual ratifications delay hold up completion of legislation for the EEC as a whole. Figure 6 represented the situation at 21 September 1988, when only eight countries had ratified. The essentiality of EEC members becoming Party is immediately apparent.

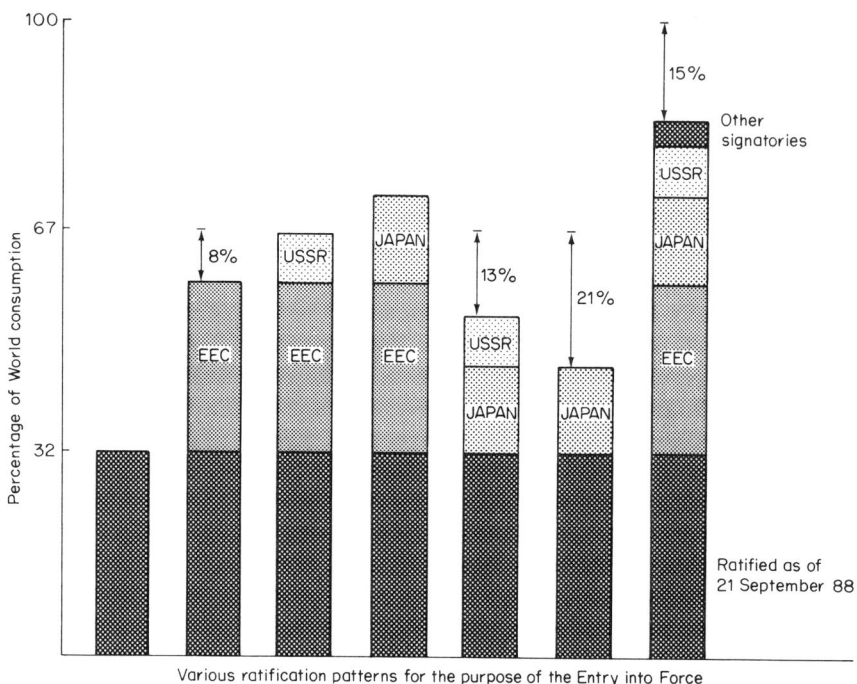

Figure 6 Global consumption of CFCs and halons controlled by the Montreal Protocol. (Source: UNEP.)

Article 8: Non-compliance

The ways and means of dealing with non-compliers was postponed until the first meeting of the Parties, following a lack of consensus at Montreal. A mechanism to safeguard against false data reporting and detection of non-compliance with agreed control measures is likely to be a complicated bureaucratic process.

Article 9: Research, development, public awareness and exchange of information

In a similar way to that specified in the Vienna Convention this article promotes cooperation in scientific research, monitoring and information exchange, particularly with regard to limiting emissions through recovery and other means

– and through substitutes. The Article requires Parties to report a summary of their activities in this regard at specified intervals.

Article 10: Technical assistance

This article reiterates the provisions of Article 4 in promoting technical assistance to developing country Parties.

Article 11: Meetings of the parties

The timetable for ordinary meetings and the provisions for calling extraordinary meetings are spelled out. The agenda for the first meeting identifying ten specific topics is included. The status of observer to meetings of the Parties can be conferred on non-party States, the United Nations and its specialized Agencies, and most other relevant bodies whether national, international, governmental or non-governmental.

Article 12: Secretariat

The duties of a secretariat to the Protocol are outlined in this Article. In practice, the secretariat for the Convention and the Protocol will be identical, although parties to the Convention need not necessarily be Parties to its Protocol. The reverse of course does not apply.

Article 13: Financial provisions

The Parties will be required to adopt financial rules and accept the costs of administrating the Protocol in accordance with standard UN procedures.

Article 14: Relationship of this Protocol to the Convention

This confirms that the provisions of the Convention apply to the Protocol except where expressly stated.

Article 15: Signature

The procedures for signature are outlined. As of 22 September 1988, 44 countries had signed. Ratification, the process of legal obligation after entry into force, has been effected by ten states.

Article 16: Entry into force

The necessary requirements for entry into force have already been discussed. As of 30 October 1988, the ten ratifications ensured that about three quarters of the second condition for entry into force (67% of global production) had been fulfilled. Ratification by the European Economic Community and by its ten member states would constitute ten, not 11, additional acceptances of the treaty. The date of entry into force is specified as 1 January 1989 provided the two conditions referred to earlier are fulfilled. If by that date they are not, then a further 90 days will have to pass beyond the date when sufficient ratifications are received. The first date after 1 January 1989 for coming into force, should insufficient ratifications occur, is 31 March 1989.

Article 17: Parties joining after entry into force

Late entrants will be required to fulfil the sum of the obligations specified in the control measures. This clause is an incentive to join early, if only to avoid the possible shock of precipitous rather than gradual reductions in consumption and production as well as being subjected to trade sanctions prior to becoming a Party.

Article 18: Reservations

The Article assures that no country can enjoy the status of a Party yet attempt to exclude itself from some obligations through tabling of reservations.

Article 19: Withdrawal

States must remain parties to the Protocol for at least 5 years before withdrawal. This clause does not apply to countries operating under Article 5, the developing countries.

Article 20: Authentic text

This informs that the Secretary General of the United Nations is the depository of the authentic texts.

Annex A: Controlled substances

This first annex lists the controlled substances in two groups. Group 1 consists of five fully halogenated chlorofluorocarbons and Group 2 three fully halogenated bromofluorocarbons. These substances are those to be controlled in accordance with Article 2. Ozone depleting potentials relative to CFC 11 are estimated for all of the compounds except halon 2402 in Group 2. It has since been estimated as six. A footnote informs of a requirement to review and revise the ODPs periodically. As noted earlier, the present list of substances can be extended or reduced by the Parties in accordance with new scientific information.

Table 2 shows the timetable Parties should adhere to in order to implement the Protocol.

Table 2 Montreal Protocol timetable

Date	Action
22 March 1985	Vienna convention adopted
16 September 1987	Montreal Protocol adopted
22 September 1988	Vienna Convention entry into force
1 January 1989	Montreal Protocol entry into force
1 April 1989	Provision of data on production, imports and exports by Parties
April 1989	First meeting of contracting Parties
1 October 1989	Assessments of scientific, environmental, economic and technical aspects completed
1 January 1990	Imports of controlled substances from non-parties banned except for Article 5 countries
April 1990	Second meeting of contracting Parties
30 June 1990	Consumption of group 1 substances frozen and production restricted to 110% of 1986 reference levels
31 December 1990	Completion of assessment of control measures by Parties
1 January 1992	Annex containing list of products containing controlled substances
1 January 1992	Report by Parties on research development, public awareness and information exchange activities
1 January 1993	Exports to non-parties cannot be subtracted in consumption calculations
1 January 1993	Exports to non-parties by Article 5 countries forbidden
31 January 1993	Consumption of group 2 substance frozen and production limited to 110% of 1986 reference levels
1 June 1993	Ban on imports of products containing CFCs included in annex elaborated for that purpose
1 January 1994	Consideration of developing an annex of substances produced with CFCs
30 June 1994	Consumption of group 1 subtracts restricted to 80% and production limited to 90% of 1986 reference levels
30 June 1994	Banning of imports of products made with controlled substances by non-objecting Parties
1 July 1999	Consumption of group 1 substances restricted to 50% and production limited to 65% of 1986 reference levels
31 June 2000	Article 5 countries freeze consumption to 0.3 kg/capita or annual average consumption for 1995–97 and limit production to 110% of reference level for this period
1 February 2003	Article 5 countries must have frozen consumption and production of group 2 chemicals to 110% of reference levels
1 July 2004	Article 5 countries must have reduced consumption of group 1 substances to 80% of reference levels
1 July 2009	Article 5 countries must have reduced consumption of group 1 substances to 50% of reference levels

THE FUTURE OF THE PROTOCOL

Assuming that joint Parties meet the obligations of the treaty and that no major group of non-parties combine to produce and use new reservoirs of the controlled substances despite the trade restrictions and denial of access to technology, the global ozone layer will, according to model calculations, stabilize at a few per cent depletion during the first decade of the next century and possibly recover thereafter. However, such model calculations are fraught with uncertainty and the report of the Ozone Trend Panel[21] has already reported observed ozone column depletions of several per cent greater than theoretical prediction, which is almost certainly attributable to ozone destruction by chlorine. These ozone losses have been particularly marked during winter in the Northern Hemisphere between 30 and 64° latitude (Figure 7).[21]

Also, with present chlorine concentrations in the stratosphere of about 2 ppbv, the Antarctic ozone hole has become a permanent and worsening feature of the

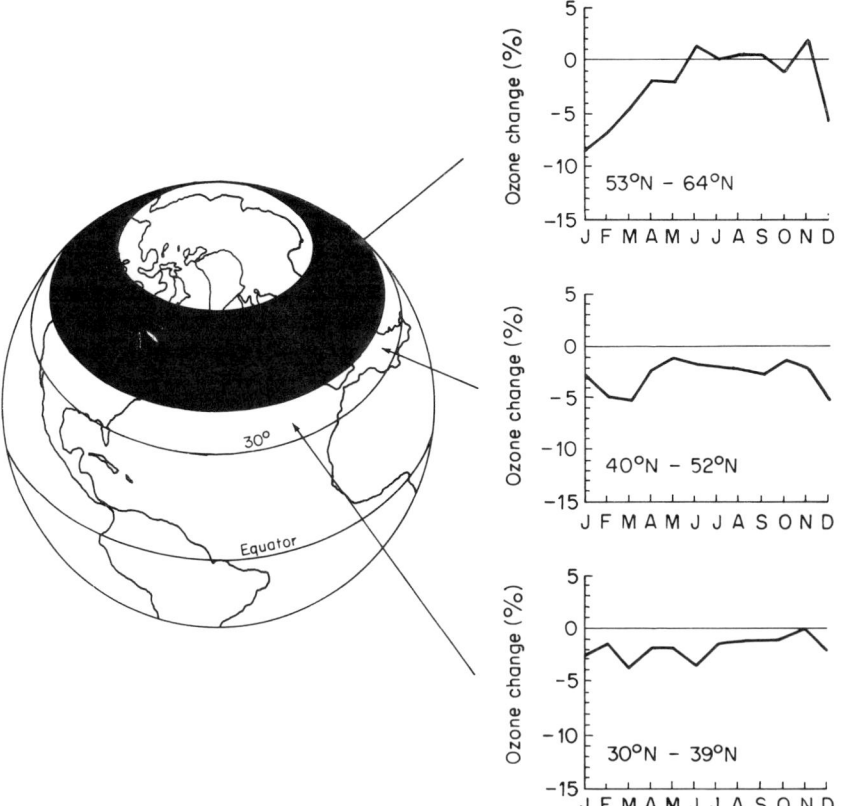

Figure 7   Ozone change 1969–86. (Source: NASA.[21])

Southern Hemisphere spring. Even with full global compliance with the Protocol, the stratospheric chlorine will increase to between 6 and 8 ppbv. It is reasonable to conclude that limiting CFC control to that required under the Protocol will maintain and probably deepen the Antarctic hole. It will require a further cutback to 85% immediately or a scheduled phase-down to 95% of reference levels by 2003 to maintain chlorine levels at 2 ppbv or less. Given the long atmospheric lifetime of the substances involved it could take decades or even centuries to reduce chlorine levels to natural values, even with a complete ban on emissions. The matter of whether or not to strengthen the Protocol will be an urgent matter for the Parties, as will the possibility of expanding the list of controlled substances in Group 1 to include methylchloroform calculated by the EPA in 1988 to account for 80% of the growth of chlorine in the atmosphere from chemicals other than the CFCs and halons. Stabilization of the ozone layer will not prevent deaths from skin cancer. Three million people will suffer that fate even with no ozone depletion. Under the implemented protocol a further 140 000 victims are expected. Without control, however, up to 5 million skin cancer deaths from 263 million skin cancer cases have been forecast (see Chapter 16).

The role of man-made chemicals in greenhouse gas warming must not be overlooked in deciding which substances should or should not be controlled under the Montreal Protocol. The greenhouse warming potential of all possible substitutes for the controlled substances and of the principal halocarbons not scheduled for control should be calculated and a clear signal given to industry on the advisability of production growth and use of them in the future.

At the World Conference on the Changing Atmosphere[22] (Toronto, May 1986) delegates cited the ultimate consequences of atmospheric pollution resulting from human activities to be second only to those of Nuclear War. The implementation of the Protocol's present and future measures is a factor in the degree of control that will eventually be necessary to address the climate change question. There can be no satisfactory solution to an environmental problem if, by reducing it, another problem is exacerbated. Thus, the role of the Protocol as a factor in climate change, acidic deposition and other atmosphere issues must constantly be reviewed by the Parties.

So serious are the demonstrated consequences of uncontrolled manipulation of the atmosphere and its composition that its protection will perhaps have to be guaranteed by a series of global agreements of which the Vienna Convention and its Montreal Protocol could be the prototype treaties.

ACKNOWLEDGEMENTS

The opinions are those of the author. Nothing in the paper necessarily represents the views of the United Nations Environment Programme, particularly with regard to the interpretation of the text of the Montreal Protocol.

The authors wishes to acknowledge the work of G. Victor Buxton of Environment Canada and that of Project Staff of OTA, whose earlier analysis of the Montreal Protocol has been used extensively as source material in the preparation of this paper.

REFERENCES

1. UNEP database 1988.
2. *Montreal Protocol on Substances that Deplete the Ozone Layer*, Final Act 1987, UNEP Na87-6106.
3. US Environmental Protection Agency. *Assessing the Risks of Trace Gases that can Modify the Atmosphere*. Washington, DC.
4. M. Molina and F. Rowland, Stratospheric sink for chlorofluoromethanes: chlorine atom-catalysed destruction of ozone. *Nature*, **249**, 810–812 (1974).
5. UNEP/GEMS Environment Library No. 2, *The Ozone Layer* (1987).
6. *UNEP Report of the Eighth Session of the Co-ordinating Committee on the Ozone Layer* (1986).
7. R. T. Watson, M. A. Geller, R. S. Stolarski and R. K. F. Hompson, *Present State of Knowledge of the Upper Atmosphere: An Assessment Report*. NASA Reference Publication 1162, May (1986).
8. UNEP Meeting of Experts designated by Governments, Inter-governmental and Non-governmental Organizations on the Ozone Layer, Washington DC, 1–9 March 1977.
9. UNEP, Ozone Layer Bulletin Nos. 1 – 9, *Assessment of Ozone Layer Modification and its Impacts*.
10. UNEP, *Report of the Governing Council on the work of its Ninth Session*, 13–26 May 1981.
11. J. Farman, B. Gardiner and J. Shanklin, Large losses of total ozone in Antarctica reveal seasonal ClO_x/NO_x interaction. *Nature*, **315**, 207–210 (1985).
12. UNEP, *Vienna Convention for the Protection of the Ozone Layer* (1985).
13. UNEP, *Report of the Governing Council on the Work of its Thirteenth Session*, 12 June 1985.
14. *Reports of the Workshops on the Control of Chlorofluorocarbons*, Rome 26–30 May 1986, Leesburg 8–12 September 1986, UNEP/WG 148.
15. *Reports of the Ad Hoc Working Group of Legal and Technical Experts for the Elaboration of a Protocol of Chlorofluorocarbons to the Vienna Convention for the Protection of the Ozone Layer*, Vienna Group, UNEP/WG 151/L4, UNEP/WG 167/2, UNEP/WG 172/2.
16. *Report of the Eighth Session of the Co-ordinating Committee on the Ozone Layer*, UNEP/CCOL/VIII and UNEP/CCOL/VIII Add. 1, Nairobi, 26–28 February 1986.
17. *Atmospheric Ozone 1985: Assessment of our Understanding of the Processes Controlling its Present Distribution and Change*. Global Ozone Research and Monitoring Project Report No. 16, WMO, Geneva (1986).
18. R. Stolarski, A. Krueger, M. Schoebel, *et al.*, Nimbus 7 satellite measurements of the springtime Antarctic ozone depletion. *Nature*, **322**, 808–811 (1986).
19. *UNEP Ad Hoc Scientific Meeting to Compose Model-Generated Assessments of Ozone Layer Change for Various Strategies for CFC Control*, Warzburg, FRG, 9–10 April 1987.
20. *UNEP Data Base for Substances to be Controlled Under the Montreal Protocol on Substances that Deplete the Ozone Layer* (1988).
21. R. T. Watson and the Ozone Trend Panel, NASA, *Present State of Knowledge of the Upper Atmosphere 1988: An Assessment Report*, June (1988).
22. *World Conference on the Changing Atmosphere*. Government of Canada, UNEP, WMO. Toronto, Ontario, June 27–30 1988.

Discussion Period 3

David Doniger (Natural Resource Defence Council)   You mention that one of the Commission members was holding out for more convincing proof, on the issue of the ozone depletion . . .

Stanley Clinton Davis   In fact it wasn't a meeting of the Commission, it was a meeting of the Council of Ministers. The Commission is entirely virtuous in this issue – it is the Ministers who are not.

Stewart Boyle (Association for the Conservation of Energy)   I am interested in the interrelationship of different directorates within the European Community. In this country there is very little cooperation between energy and environment, even though any solution to the greenhouse effect must involve both departments.

Stanley Clinton Davis   I can only say that we act as a Commission, and on the greenhouse issue we have acted in unity. We have produced a very comprehensive analysis of the options available for dealing with the greenhouse effect and all the services have cooperated towards that objective. Incidentally, I do wish that Ministers would apply themselves much more to the question of energy conservation and to the question of the more efficient use of energy rather than living in a fantasy world and talking about a massive escalation of nuclear energy. Massive escalation is not an option. It is not feasible having regard to the political feeling within the whole of the community, and the cost is not bearable. I do not say that one should exclude completely some form of nuclear development, but certainly one must reject at once the concept of a massive dependence upon nuclear energy.

Robin Russell Jones   Dr Usher, thank you very much for that detailed explanation of how the Montreal Protocol came into effect. Could I on a point of clarification just ask you to explain the differences between the fact that the

Protocol, presuming it is fully implemented, will stabilize levels of stratospheric ozone, whereas levels of chlorine will continue to increase?

Peter Usher Essentially we are talking about global ozone levels in this particular case. I was not talking about a regional ozone hole. Perhaps Bob Watson would like to comment on that?

Bob Watson Yes, I think what you said could be dangerously misleading, unfortunately. I think what you are really referring to is the two-dimensional models that John Pyle and Professor Isaksen run. These would not predict a major ozone depletion under the Montreal Protocol, the reason simply being that predicted ozone decrease from CFCs is offset by other pollutants such as carbon dioxide increases. However, this would entail a major loss of ozone at 40 km and a major increase near the Earth's surface.

Robin Russell Jones I know I have tried quite hard to interest some of the governments that are non-signatories to the Montreal Protocol to attend this conference. Do we have any representatives from Taiwan, Korea or India – No? Well I think that tells us a little bit about why we have problems.

Bob Watson Peter, it is said that the strength of the Vienna Convention relies on the exchange of information and research. Clearly this has been very valuable and has helped pave the way for the Montreal Protocol. One thing that bothers me about the talks this morning and hearing Stanley Clinton Davis this afternoon is that governments have now recognized the importance of the issue but are not willing to commit the necessary resources. Have you heard of any nation in the world that is willing to increase the level of research in order to obtain a better understanding of how climates are likely to change? Governments and UNEP are demanding more and more understanding with less and less resources. I think this is a losing situation, when all future policies are going to be based on very poor scientific understanding, and that is not a healthy situation.

Peter Usher You have put your finger on a very major problem, not only for the research community but for the international community also. UNEP does not have the staff, nor does it have the money to do all of the things that are asked of us. There is an urgent need for research now into all atmospheric problems, not just ozone, for example to confirm the trends in global warming that have been outlined today. This is probably one of the most serious issues that we are facing in the world today and yet resources are being withheld, both from the research community and from those people who are obliged to forge international agreements. I can only echo your concern.

R. Pomerance You mention that radiative effects would be included in the forthcoming scientific review. Will this include the chemicals that are not currently controlled by the Montreal Protocol? Today we have had an interesting, perhaps one of the first public discussions of the greenhouse effect of some of these substitutes. They appear to be significant.

Peter Usher Perhaps I was a little premature in putting that on the screen or suggesting that the process is finalized. The purpose of the meeting in London this week will be to determine exactly what is to be reviewed. There will be representatives from all the major research countries present. It will be headed by Bob Watson and I am sure that he will be taking all the relevant information into account. I cannot say more than that.

Robin Russell Jones Virginia Bottomley and Stanley Clinton Davis both talked about a CFC-free world. Could I just ask you very simply what obstacles you see to realizing that situation and what initiatives are needed to make sure that they are overcome?

Peter Usher I think the major problem that we are facing is one that you alluded to yourself – that there are many countries that are not on board the Protocol. At the moment we have 35 signatories. They do not include the developing countries with large populations and the capacity to make these chemicals. It is just possible that new large producers and users of these chemicals could emerge and stay outside of the Protocol. The Protocol itself is extremely thorough. It has trade restrictions between Parties and non-parties and for non-parties to stay outside the Protocol then, under the terms of the Protocol, they are denied technology and they are denied trading partners among the parties to the Protocol. However, even staying within the limitations of the Protocol, signatory countries are allowed 0.3 kg/person per year. If the non-producers reached that limit they could double the production of CFCs worldwide.

Now that is an enormous concession. It is hoped that in practice this would never happen because of the move away from the regulated substances towards new technology and ozone-friendly substitutes. But certainly there are countries large enough to upset the treaties that have been agreed by others. It is imperative that the international community broadcasts the importance of this issue. Often one is told by the developing countries that these are problems for the developed world; that they created them, not us. Our problems are more immediate than this. We are dying of drought and so we cannot be primarily concerned with long-term problems. Yet I think it is important for everyone to recognize that this is a global problem. No country is immune from the environmental and climatic changes that will result from atmospheric pollution.

Part 5
Ultraviolet-Induced Carcinogenesis

12
Photosensitive Human Syndromes and Cellular Defects in DNA Repair

C. F. ARLETT and J. COLE
MRC Cell Mutation Unit, University of Sussex, UK

The wavelengths of light reaching the Earth's surface have been monitored carefully[1] and indicate that the ozone layer absorbs all UVC wavelengths and a significant part of the UVB region. Thus a reduction in ozone levels will inevitably lead to an increase in the exposure of human beings to UVB radiation. The action spectrum for the induction of DNA damage in the form of pyrimidine dimers,[2] a lesion known to be of major significance for lethality, mutation and cancer,[3] shows that, while UVC is most efficient at inducing this lesion, UVB is also effective.

Health effects as a consequence of solar radiation may be divided into three categories:

(1) Skin damage including non-melanoma skin cancer and melanoma.
(2) Ocular abnormalities including cataract and retinal changes.
(3) Changes in the immune system, which may alter the course of cancer or infectious disease.[4]

The genodermatoses, in particular xeroderma pigmentosum (XP), are exquisitely sensitive to solar irradiation[5] and thus provide us with an accelerated model of the situation that might obtain with normal individuals given that if more UVB impinges on the Earth's surface more DNA damage must be

---

Ozone Depletion: Health and Environmental Consequences
Edited by R. Russell Jones and T. Wigley
© 1989 John Wiley & Sons Ltd

sustained and expressed. Most cellular studies report on the response to UVC at 254 nm, close to the absorption maximum of nucleic acid. These results are likely to give a more dramatic representation of what will be seen following more substantial doses of UVB.

The severity of the clinical manifestations of XP are directly related to the amounts of sunlight received by the individual and are seen to be most severe in countries receiving greater insolation such as Egypt.[6] These unfortunate individuals can be given protection in the form of sunscreens and adequate clothing and by encouraging a lifestyle with minimal exposure to sunlight.[5] Perhaps this is the message that should be given to any population of normal individuals who might become exposed to more UVB, either electively or as a consequence of a reduction in the ozone layer.

XERODERMA PIGMENTOSUM

Clinical observations

Xeroderma pigmentosum is an autosomal recessive disease characterized by sun-sensitivity coupled with a distinctive erythemal response.[6] Kraemer et al.[7] have reviewed 830 published cases and these data provide a comprehensive picture of the disease.

The dermatological effects are seen as pigmented macules, achromic spots and telangiectasia in sun-exposed areas followed, ultimately, by basal cell carcinoma (BCC), squamous cell carcinoma (SCC) and malignant melanoma. Individuals with the disease die about 30 years earlier than the US general population. The median age of onset of cutaneous symptoms was 1–2 years and the onset of symptoms was delayed until after 14 years of age in 5% of patients. Of patients in the 10–14 year age group, 50% had skin cancers and the median age for first skin neoplasms is roughly 50 years earlier than for the US population in general.

The role of sunlight in the induction of cancer in these patients is unambiguous. Thus, the distribution of 97% of the BCC plus SCC is in that region of the body (the face, head and neck) that receives most direct sunlight. This is in contrast with a figure of 80% for the US general population. The distribution of melanomas in XP patients is different from that in the normal population; 35% occurred on sites other than face, head or neck, whereas 80% occurred at the other sites in the US general population. The role of sunlight in the aetiology of melanomas is complex, but the incidence of cutaneous changes (including melanomas) related directly to exposure to sunlight appears irrefutable in XP. A 10 to 20-fold increase in the frequency of internal neoplasms derived from registry data has also been claimed for XP.[8]

Further support for the involvement of sunlight in the expression of the XP phenotype is seen in the tissues of the eye, including the lids, conjunctiva and the cornea and the oral region. The response of the lids is, as expected, characteristic of XP skin. Photophobia and conjunctival infections are common, as are destructive changes in the cornea. The frequency of ocular cancers has been estimated as 2×10^3 times elevated in XP patients under the age of 20 years. The frequency of SCC of the tip of the tongue is 2×10^4 times elevated.

Neurological defects were described in 20% of patients. Here, of course, no direct involvement of sunlight can be inferred. An early description of three siblings with the cutaneous features of the disease coupled with microcephaly, progressive neurological degeneration, dwarfism and impaired sexual development led to the definition of the De Sanctis–Cacchione syndrome.[9] The median age was 6 months for the onset of skin symptoms for patients with neurological abnormalities, compared with 2 years for XP patients with no such abnormalities. A small proportion (5%) of the individuals with neurological defects showed the defects after the age of 5 years.

A number of other clinical abnormalities have been associated with XP.[5] It is of some interest to note that impaired immune status has been reported in the disease. This includes reduced responses to recall antigens and DNCB antigens, a reduction in the ratio of circulating T helper/suppressor cells, and a reduced response to phytohaemagglutinin stimulation of lymphocytes.[10-12] There is, however, considerable variation among patients with regard to these effects. For example, the extent of reduced DNCB sensitivity was shown[11] to correlate with the extent of the severity of the disease, which in turn probably reflected the extent of exposure in the series of patients examined. Very recently Norris et al. (personal communication) have examined immune function in five young Caucasian children with XP who had been well protected from sunlight. Cutaneous hypersensitivity was normal, as was the lymphocyte proliferation response. The levels of natural killer (NK) cell activity was greatly reduced, but the number of circulating cells of the relevant class was normal, indicating a defect in NK function.

Cellular studies

The clinical features of XP are a good indicator of the effects of sunlight on these patients. Are there any cellular responses that may assist in our understanding of the disease? The first indication of a cellular defect was recorded by Gartler,[13] who reported hypersensitivity to UVC. Cleaver[14] demonstrated a defect in excision repair in XP fibroblasts, and this was followed shortly by reports of the so-called 'variant' form of XP, which is competent in excision repair.[15] Variant fibroblasts are defective in daughter strand repair[16] and are also minimally sensitive to the lethal effects of 254 nm ultraviolet light[15,17] (Figure 1).

Keyse et al.[18] showed that the greater sensitivity of excision-defective XP cells to 254 nm light compared with normal fibroblasts decreases with increasing wavelength, suggesting different mechanisms for inactivation at longer wavelengths. Hypersensitivity to the lethal action of DNA-damaging agents such

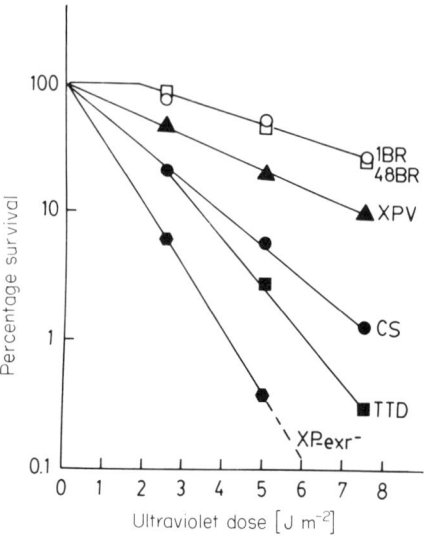

Figure 1 The lethal effects of 254 nm ultraviolet light on dermal derived fibroblasts. 1BR and 48BR, normals; XPV, XP variant (patient AM); CS, Cockayne's syndrome; TTD, trichothiodystrophy cell culture TTD1BR; XP-exr⁻, excision defective xeroderma pigmentosum.

as benz(a)pyrene or 4-nitroquinoline-1-oxide, which induce bulky adducts on DNA has also been reported.[19,20]

The defects in excision repair have been used as the basis of a complementation test and, to date, nine excision-defective complementation groups, A-I, have been assigned in the disease.[21] The variants represent a single group of approximately 20% of cases. There is an uneven distribution of complementation groups on a worldwide basis.[7,22] Groups C, D, A and variant are the most common in Europe and the USA, while in Japan groups A and variant are most common and group C is rarely encountered. Complementation groups B, H, and I are rare, being represented by single kindreds. The single individuals belonging to groups B and H are unique in that they also have the symptoms of Cockayne's syndrome.[23,24] The overlap between syndromes is also seen between group D and trichothiodystrophy.[25]

The defects in repair are sufficiently robust that they may be used to confirm clinical diagnosis and have been employed with success to perform prenatal diagnosis in families with one affected child.[26,27]

Mutation studies utilizing 254 nm ultraviolet-induced resistance to 6-thioguanine, 8-azaguanine or diphtheria toxin have revealed that XP fibroblasts, both excision-defective and excision-proficient, are hypermutable when compared with cells from normal individuals.[28-30] Data for 6-thioguanine resistance

Table 1 Induction of mutants resistant to 2.5 µg/ml 6-thioguanine by 254 nm ultraviolet light in cultured human fibroblasts

Donor	Type	Mutant frequency per survivor × 10⁻⁶		
		Spontaneous	1 J/m²	3 J/m²
Normal				
GM730		10.0	12.0	16.0
2BI		2.5	6.5	14.0
1BR		4.0	3.0	8.0
54BR		0.8	1.0	4.0
Xeroderma pigmentosum				
XP4LO	Group A	4.0	120.0	—
XP1BR	Group D	4.5	48.0	—
XP2BI	Group G	4.6	47.0	—
XP3BR	Group G	5.5	38.0	—
XP6DU	Variant	0.8	4.0	60.0
XP30RO	Variant	2.2	10.0	40.0
Cockayne's syndrome				
CS697TO		6.9	14.0	50.0
Trichothiodystrophy				
TTD1GL		2.5	4.0	10.8
TTD2GL		1.0	42.0	—
TTD1BI		4.0	6.0	15.0

are shown in Table 1. Xeroderma pigmentosum cells are also hypersensitive to the induction of sister chromatid exchanges (SCE)[31] and chromosome aberrations by UVC.[32] The elegant studies of Maher et al.[33] utilizing a technique for the repair of damage under conditions in which cells are taken out of the division cycle reveal that (at least under these conditions) the excision repair process is essentially error-free. We might therefore assume that it is the daughter strand repair process which, under greater stress in the excision defective cells,[16] is the error-prone process generating enhanced mutation.

Recently, a technique has become available that permits the measurement of the frequency of 6-thioguanine-resistant mutants in circulating T lymphocytes from samples of peripheral blood.[34] Results of such studies with lymphocytes from XP patients show elevated mutant frequencies (Table 2). At first sight, one may wonder why lymphocytes should show elevated mutant frequencies. When we remember that all the lympocytes circulate through the skin in a period of approximately 10 min, an explanation based on induction rather than an elevated intrinsic mutation rate seems most plausible. It is clear that older patients who were not protected as children show a much greater enhancement compared with normals than the well protected younger patients. We have also been able to compare the sensitivities of fibroblasts and T lymphocytes derived from the same

Table 2 Mutant frequency to 5×10^{-6} M 6-thioguanine resistance in circulating T lymphocytes

Donor	Age (years)	Mutant frequency $\times 10^{-6}$	
		1 SD	1 SEM
Children ($n=13$)	1–17	2.93 ± 2.44	(0.68)
Normal adults			
Non-smokers ($n=20$)	22–52	5.04 ± 2.23	(0.50)
Smokers ($n=20$)	19–67	10.82 ± 7.92	(± 1.87)
XP			
1	8	4.66	
2	13	3.58	
3	16	7.11	
4	13	29.72	
5	16	10.86	
6	10	5.0	
7	34	28.2	
variant 1	48	32.0	
variant 2	62	19.9	
CS			
1	2	1.4	
2	2	1.86	
3	14	7.92	
4	14	7.28	
TTD 1 (TT1BR)	15	<1.02	

SD, standard deviation; SEM, standard error of the mean.

patients, and the results reveal that although the differences between XP and normal are maintained, lymphocytes in general are more sensitive to UVC than fibroblasts (Figure 2). The sensitivity of immune functions remains unknown.

Inevitably XP has been a fertile source of material for the study of DNA repair. The injection of phage T4 endonuclease[35] can restore excision repair to some XP cells and reduce their hypersensitivity to the lethal action of ultraviolet light. The use of cell-free extracts by Mortelmans et al.[36] showed that XP cells were able to excise pyrimidine dimers from DNA but not from DNA combined with protein, indicating that the defects are associated with factors that make the damage accessible to the ultraviolet endonuclease. Unscheduled DNA synthesis was restored to cells of nine complementation groups after microinjection of Micrococcus luteus ultraviolet endonuclease.[37] Cells from XP complementation group E (two cases only) have been found to lack a factor that binds to ultraviolet damaged DNA,[38] emphasizing the possible importance of accessibility factors.

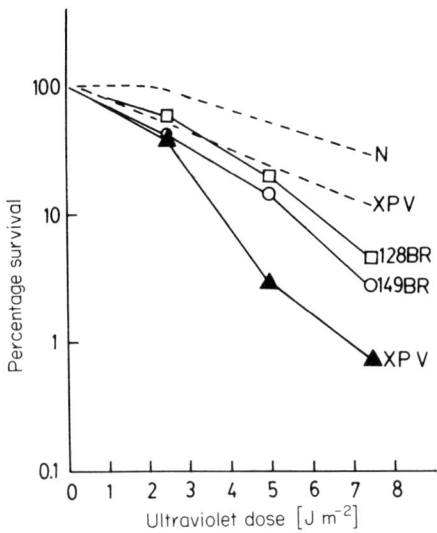

Figure 2 The lethal effect of 254 nm ultraviolet light on T lymphocytes. The dotted lines are taken from 1BR and 48BR the normal and XPV variant (patient AM) fibroblasts from Figure 1. 128BR and 149BR, normals; XPV, XP variant (patient AM).

However, it was pointed out by Kaufmann[39] that attempts to achieve *in vitro* complementation of the XP phenotype has led to many contradictory results. For example, the interesting results of Wood *et al.*[40] gave indications of a defect in excision repair in the excision-proficient XP variants. To date, attempts to provide a convincing chromosome location for the XP gene (s) have not been successful. It remains to be seen whether the diversity of complementation groups will be resolved by mapping. A human 17 kb gene that corrects a defect in excision repair in Chinese hamster cells has been cloned and sequenced[41] and this approach may also be applicable to the study of defects in human DNA repair. The ultraviolet sensitivity of an XP cell line has been successfully corrected following a DNA transfection experiment of heroic proportions.[42] The eventual cloning and sequence of the XP gene is to be anticipated. One difficulty that still remains in these studies is the possibility that wild-type revertants may arise during multiple exposures to UVC as a consequence of the hypermutability of these cells.[43]

Xeroderma pigmentosum cells have also been used as a substrate in mutation studies with shuttle vectors, and the results[44] show that plasmid survival is lower in XP cells than in repair competent cells. The mutations scored in the vector PZ189 showed a hundred-fold increase in frequency in the marker gene, and subsequent analysis revealed that the mutation spectrum was altered when the plasmid was passaged in XP cells. More GC to AT transitions and fewer transversions GC to TA were observed.

The salient features of XP that are relevant to any study of ultraviolet-induced carcinogenesis in man concern the correlations between skin carcinogenesis, defects in DNA repair and enhanced mutation when cells are exposed to ultraviolet light. The existence of this data set provides proof of the mutational theory of cancer induction, and it has become axiomatic that defects in DNA repair are important, but do these assumptions respond to critical analysis?

The first and most obvious problem relates to the fact that the mutation observations have been made with UVC, which is not present in the solar spectrum that reaches the Earth's surface. However, Patton et al.[45] have confirmed hypermutability using wavelengths simulating sunlight. Investigations of two other genophotodermatoses reveal that the relationship between skin cancer, DNA repair and enhanced mutation is more complex than suggested by the study of XP alone.

COCKAYNE'S SYNDROME

Clinical observations

Cockayne's Syndrome (CS) is a rare syndrome exhibiting a pattern of inheritance consistent with it being recessive.[46] Severe photosensitivity is associated with dwarfism, mental retardation, and skeletal and retinal defects. Loss of adipose tissue leading to an ageing effect is noted, together with neurological degeneration.[47] In marked contrast to XP, there are no reports of increased incidence of cancer. The individuals have none of the characteristic skin changes shown by XP, and death has usually occurred by 20 years of age. Two individuals who showed XP-like skin changes were shown to suffer from both XP and CS.[23,24]

Cellular studies

Cockayne's syndrome cells are hypersensitive to the lethal effects of UVC (Figure 1) although this is, in general, not so marked as in excision-defective XP cells. Sensitivities to other DNA-damaging agents are broadly similar to those of XP.[47] The cells of CS are also hypersensitive to the induction of SCE.[48] Both excision repair and daughter strand repair are normal in this condition. Cells from CS patients may be distinguished from normals by the failure of DNA and RNA synthesis to recover from ultraviolet irradiation.[49,50] These distinctive defects have allowed the assignment of at least three complementation groups in the syndrome.[51] The defect in recovery of post-ultraviolet RNA synthesis has been used as the basis of prenatal diagnosis.[52] The defect has been interpreted as implying possible defects in processing on damaged templates,[50] and recently[53] these cells have been shown to carry out preferential repair of transcribing DNA.

Studies of mutation have been hampered by the fact that fibroblast cultures of CS tend to age rapidly in culture, presenting severe technical limitations.[29] However, successful experiments with one culture (Table 1) reveal them to be at least as mutable as XP variant cells. This observation is supported by studies of mutation induction of herpes simplex virus when grown in lymphoblastoid cultures of CS.[54]

The measurement of mutant frequencies of 6-thioguanine-resistant cells amongst the T lymphocytes of CS patients (Table 2) reveals a two-fold enhancement in two older children (14 years old) compared with normals, but no enhancement in the two younger children (2 years old).

TRICHOTHIODYSTROPHY

Clinical observations

Trichothiodystrophy (TTD) is a rare syndrome having a mode of inheritance consistent with a recessive gene.[47,55] The unifying feature is the presence of sulphur-deficient brittle hair associated with a marked reduction in the content of the sulphur-containing amino acids cysteine and cystine. Mental and physical retardation is variable. An unusual facies is common, which can be reminiscent of CS. Ichthyosis is also common, and severe photosensitivity has been reported in about half the patients. No XP-like changes in skin or cancer have been reported for these very rare patients.

Cellular studies

Investigations with fibroblast cultures established from these patients have given complex results. We have recently reported on a detailed investigation with three unrelated patients who were not considered photosensitive.[56] Cells from Patient 1 (TTD1GL) were indistinguishable from normal. Cells from Patient 2 (TTD2GL) were indistinguishable from those of excision defective XP and, in agreement with a previous study, were shown to be in complementation group D of XP.[25] The cells from Patient 3 (TTD1BI) were unique in having a 50% deficit in excision repair, a 30% deficit in post-ultraviolet RNA synthesis but normal sensitivity to the lethal and mutagenic (Table 1) effects of UVC. These cells are again assigned to complementation group D of XP. We have been able to make only one set of observations on mutant lymphocyte frequency in a single (photosensitive) TTD patient (TTD1BR). This patient has the cellular characteristics of Patient 2 above (see Figure 1 for survival data); no mutants were detected.

SYNTHESIS

The early onset of cancer in XP[5] provides a useful model for the consequences of increased solar damage to DNA in normal individuals. The increased exposure to

UVB could be caused either intentionally from a lifestyle that encourages sunbathing or unintentionally as a consequence of ozone depletion. Individuals suffering from XP may be protected to a large extent from the dermal effects of UVB.[5] It is of some interest to note that in the series of XP patients who were tested for the frequency of mutant T lymphocytes circulating in the blood, most of those younger individuals who had been diagnosed earlier and managed more rigorously had less elevated mutant frequencies compared with age-matched controls than the older patients (Table 2). It is also possible to achieve a degree of protection against premalignant and malignant skin lesions with isotretinoin therapy.[57] This protection is reversed rapidly when therapy is withdrawn. Thus amelioration of the consequences of increased UVB damage in XP carries with it the lesson for normal individuals that changes in lifestyle are to be recommended. It also holds out the remote prospect of therapy for normal individuals if exposed unintentionally to more UVB.

A second important implication of the study of XP has been the support the increased mutability has given to the concept of a mutational origin of cancer. The defects in DNA repair observed in XP clearly have implications for cancer induction, since in addition to increased skin cancer these patients have been also reported to have an increased incidence of some internal tumours.[8] Thus proficient DNA repair protects against cancer. However, the absence of cancer in both CS and TTD, despite increased induced mutability in both CS and some TTD, and decreased DNA repair in some TTD, suggests that there must be an additional defect in XP. Bridges[58] pointed out that XP might be defective in immune surveillance. Observations on this have been conflicting.[10–12] Very recently (P. Norris, St Thomas' Hospital, London, personal communication), in a comparative study of XP, CS and a single TTD patient, defects in NK cell function (but not number) were recorded for XP alone. While it is difficult to envisage a role for NK cells in the ontogeny of skin cancer the suppression of a component of the immune system in these patients draws our attention to the importance of immune suppression following exposure to UVB.[4,59] The increased sensitivity of T lymphocytes compared with fibroblasts (Figure 2) to the lethal effects of ultraviolet light may also be of significance for effects on the immune system.

ACKNOWLEDGEMENTS

We are indebted to S. A. Harcourt and G. Stephens for their expert technical assistance and to Professor B. A. Bridges for his helpful comments. Results reported in this paper were supported by EEC Contract EV4V–0037UK.

REFERENCES

1. I. Magnus, *Dermatological Photobiology, Clinical and Experimental Aspects*. Blackwells, Oxford (1976).

2. R. Setlow, The wavelengths in sunlight effective in producing skin cancer: a theoretical analysis. *Proc. Natl. Acad. Sci. USA*, 77, 3363–3366 (1974).
3. E. C. Friedberg, *DNA Repair*. Freeman, San Francisco (1985).
4. M. Giannini, Suppression of pathogenesis in cutaneous leishmaniasis by UV irradiation. *Infect. Immun.* 51, 838–845 (1986).
5. K. Kraemer, Xeroderma pigmentosum. In D. J. Demis, R. L. Dobson and J. McGuire, eds, *Clinical Dermatology*, Harper and Row, Hagerstown, Vol. 4, pp 1–33 (1980).
6. C. Ramsay and F. Giannelli, The erythemal action spectrum and deoxyribonucleic acid repair synthesis in xeroderma pigmentosum. *Br. J. Dermatol.*, 92, 49–56 (1975).
7. K. Kraemer, M. Lee and J. Scotto, Xeroderma pigmentosum: cutaneous, ocular and neurologic abnormalities in 830 published cases. *Arch. Dermatol.*, 123, 241–250 (1987).
8. K. Kraemer, M. Lee and K. Scotto, DNA repair protects against cutaneous and internal neoplasia: evidence from xeroderma pigmentosum. *Carcinogenesis*, 5, 511–514 (1984).
9. W. Reed, G. Sugarman and R. Mathias, De Sanctis–Cacchione syndrome. *Arch. Dermatol.*, 113, 1561–1563 (1977).
10. J. Dupuy and D. Lafforet, A defect of cellular immunity in xeroderma pigmentosum. *Clin. Immunol. Immunopathol.*, 3, 52–58 (1974).
11. W. Morison, C. Bucana, N. Hashem, M. Kripke, J. Cleaver and J. German, Impaired immune function in patients with xeroderma pigmentosum. *Cancer Res.*, 45, 3929–3931 (1985).
12. A. Wysenbeek, H. Weiss, M. Duczyminer-Kahana, M. Grunwald and A. Pick, Immunologic alterations in xeroderma pigmentosum patients. *Cancer*, 58, 219–221 (1986).
13. S. Gartler, Inborn errors of metabolism at the cell culture level. In M. Fishbein, ed., *Second International Conference on Congenital Malformations*, International Medical Congress, New York, p. 94 (1964).
14. J. Cleaver, Deficiency in repair replication of DNA in xeroderma pigmentosum. *Nature*, 218, 652 (1968).
15. J. Cleaver, Xeroderma pigmentosum: variants with normal DNA repair and normal sensitivity to UV light. *J. Invest. Dermatol.*, 58, 124–128 (1972).
16. A. Lehmann, S. Kirk-Bell, C. Arlett *et al.*, Xeroderma pigmentosum cells with normal levels of excision repair have a defect in DNA synthesis after UV-irradiation. *Proc. Natl. Acad. Sci. USA*, 72, 219–223 (1975).
17. C. Arlett, S. Harcourt and B. Broughton, The influence of caffeine on cell survival in excision-proficient and excision-deficient xeroderma pigmentosum and normal human cell strains following ultraviolet light irradiation. *Mut. Res.*, 33, 341–346 (1975).
18. S. Keyse, S. Moss and D. Davies, Action spectra for inactivation of normal and xeroderma pigmentosum human skin fibroblasts by ultraviolet radiations. *Photochem. Photobiol.*, 37, 307–312 (1983).
19. V. Maher, J. McCormick, P. Grover and P. Sims, Effect of DNA on the cytotoxicity and mutagenicity of polycyclic hydrocarbon derivatives in normal and xeroderma pigmentosum human fibroblasts. *Mut. Res.*, 43, 117–138 (1977).
20. J. Cleaver and D. Bootsma, Xeroderma pigmentosum: biochemical and genetic characteristics. *Ann. Rev. Genet.*, 9, 19–38 (1975).
21. E. Fischer, W. Keijzer, H. Thielmann *et al.*, A ninth complementation group in xeroderma pigmentosum, XP I. *Mut. Res.*, 145, 217–225 (1985).
22. H. Takebe, Y. Miki, T. Kozuka *et al.*, DNA repair characteristics and skin cancers of xeroderma pigmentosum patients in Japan. *Cancer Res.*, 37, 490–495 (1977).

23. J. Robbins, K. Kraemer, M. Lutzner, B. Festoff and H. Coon, Xeroderma pigmentosum, an inherited disease with sun sensitivity, multiple cutaneous neoplasms, and abnormal DNA repair. *Ann. Intern. Med.*, **80**, 221–248 (1974).
24. A. Moshell, M. Ganges, M. Lutzner *et al.*, A new patient with both xeroderma pigmentosum and Cockayne syndrome establishes the new xeroderma pigmentosum complementation group H. In E. C. Friedberg and B. A. Bridges, eds, *Cellular Responses to DNA Damage*, Alan R. Liss, New York, pp 209–213 (1983).
25. M. Stefanini, P. Lagomarsini, C. Arlett *et al.*, Xeroderma pigmentosum (complementation group D) mutation is present in patients affected by trichothiodystrophy with photosensitivity. *Hum. Genet.*, **74**, 107–112 (1986).
26. D. Halley, W. Keijzer, N. Jaspers *et al.*, Prenatal diagnosis of xeroderma pigmentosum (group C) using assays of unscheduled DNA synthesis and postreplication repair. *Clin. Genet.*, **16**, 137–146 (1979).
27. C. Ramsay, T. Coltart, S. Blunt, S. Pawsey and F. Giannelli, Prenatal diagnosis of xeroderma pigmentosum. *Lancet*, **i**, 1109–1112 (1974).
28. V. Maher and J. McCormick, Effect of DNA repair on the cytotoxicity and mutagenicity of UV irradiation and of chemical carcinogens in normal and xeroderma pigmentosum cells. In J. M. Yuhas, R. W. Tennant and J. B. Regan, eds, *Biology of Radiation Carcinogenesis*, Raven Press, New York, pp 129–145 (1976).
29. C. Arlett and S. Harcourt, Variation in response to mutagens amongst normal and repair-defective human cells. In C. W. Lawrence, ed., *Induced Mutagenesis Molecular Mechanisms and their Implications for Environmental Protection*, Plenum Press, New York, pp 249–266 (1982).
30. T. Glover, C. C. Chang, J. Trosko and S. S. I. Li, Ultraviolet light induction of diphtheria toxin-resistant mutants in normal and xeroderma pigmentosum human fibroblasts. *Proc. Nat. Acad. Sci. USA*, **76**, 3982–3986 (1979).
31. E. De Weerd-Kastelein, W. Keijzer, G. Rainaldi and D. Bootsma, Induction of sister chromatid exchanges in xeroderma pigmentosum cells after exposure to ultraviolet light. *Mut. Res.*, **45**, 253–261 (1977).
32. R. Marshall and D. Scott, The relationship between chromosome damage and cell killing in UV irradiated normal and xeroderma pigmentosum cells. *Mut. Res*, **36**, 397–400 (1976).
33. V. Maher, D. Dorney, A. Mendrala, B. Konze-Thomas and J. McCormick, DNA excision repair processes in human cells can eliminate the cytotoxic and mutagenic consequences of UV irradiation. *Mut. Res*, **62**, 311–323 (1979).
34. J. Cole, M. Green, S. James, L. Henderson and H. Cole, Human population monitoring: a further assessment of factors influencing measurements of thioguanine-resistant mutant frequency in circulating T-lymphocytes. *Mut. Res.*, **204**, 493–507 (1988).
35. K. Tanaka, M. Sekiguchi and Y. Okada, Restoration of ultraviolet-induced unscheduled DNA synthesis of xeroderma pigmentosum cells by the concomitant treatment with bacteriophage T4 endonuclease V and HVJ (Sendai virus). *Proc. Nat. Acad. Sci. USA*, **72**, 4071–4075 (1975).
36. K. Mortelmans, E. Friedberg, H. Slor, G. Thomas and J. Cleaver, Defective dimer excision by cell-free extracts of xeroderma pigmentosum cells. *Proc. Natl. Acad. Sci. USA*, **8**, 2757–2761 (1976).
37. A. De Jonge, W. Vermeulen, B. Klein and J. Hoeijmakers, Microinjection of human cell extracts corrects xeroderma pigmentosum defect. *EMBO J*, **2**, 637–641 (1983).
38. G. Chu and E. Chang, Xeroderma pigmentosum group E cells lack a nuclear factor that binds to damaged DNA. *Science*, **242**, 564–567 (1988).

39. W. Kaufmann, *In vitro* complementation of xeroderma pigmentosum. *Mutagenesis*, **3**, 373–380 (1988).
40. R. Wood, P. Robins and T. Lindahl, Complementation of the xeroderma pigmentosum DNA repair defect in cell-free extracts. *Cell*, **53**, 97–106 (1988).
41. A. Westerveld, J. Hoeijmakers, M. Van Duin *et al.*, Molecular cloning of a human DNA repair gene. *Nature*, **310**, 425–429 (1984).
42. K. Tanaka, I. Satokata, Z. Ogita and Y. Okada, Toward the molecular cloning of the gene for xeroderma pigmentosum, complementation group A by DNA transfection method. *J. Cell. Biochem.*, Suppl **12A**, 246 (1988).
43. B. Royer-Pokora and W. Haseltine, Isolation of UV-resistant revertants from a xeroderma pigmentosum complementation group A cell line. *Nature*, **311**, 390–392 (1984).
44. A. Bredberg, K. Kraemer and M. Seidman, Restricted ultraviolet mutational spectrum in a shuttle vector propagated in xeroderma pigmentosum cells. *Proc. Natl. Acad. Sci. USA*, **83**, 8273–8277 (1986).
45. J. Patton, L. Rowan, A. Mendrala, J. Howell, V. Maher and J. McCormick, Xeroderma pigmentosum fibroblasts including cells from XP variants are abnormally sensitive to the mutagenic and cytotoxic action of broad-spectrum simulated sunlight. *Photochem. Photobiol.*, **39**, 37–42 (1984).
46. V. McKusick, *Mendelian Inheritance in Man*, The Johns Hopkins University Press, Baltimore (1986).
47. A. Lehmann, Cockayne's syndrome and trichothiodystrophy: defective repair without cancer. *Cancer Rev.*, **7**, 82–103 (1987).
48. R. Marshall, C. Arlett, S. Harcourt and B. Broughton, Increased sensitivity of cell strains from Cockayne's syndrome to sister-chromatid-exchange induction and cell killing by UV light. *Mut. Res.*, **69**, 107–112 (1980).
49. A. Lehmann, S. Kirk-Bell and L. Mayne, Abnormal kinetics of DNA synthesis in ultraviolet light-irradiated cells from patients with Cockayne's syndrome. *Cancer Res.*, **39**, 4237–4241 (1979).
50. L. Mayne and A. Lehmann, Failure of RNA synthesis to recover after UV-irradiation: an early defect in cells from individuals with Cockayne's syndrome and xeroderma pigmentosum. *Cancer Res.*, **42**, 1473–1478 (1982).
51. A. Lehmann, Three complementation groups in Cockayne syndrome. *Mut. Res.*, **106**, 347–356 (1982).
52. A. Lehmann, A. Francis and F. Giannelli, Prenatal diagnosis of Cockayne's syndrome. *Lancet*, **i**, 486–488 (1985).
53. L. Mayne, L. Mullenders and A. Van Zeeland, Cockayne's syndrome: a UV sensitive disorder with a defect in the repair of transcribing DNA but normal overall excision repair. In E. Friedberg and P. Hanawalt, eds, *Mechanisms and Consequences of DNA Damage Processing*, Alan R. Liss, New York pp 349–353 (1988).
54. E. Henderson and W. K. Long, Host cell reactivation of UV- and X-ray-damaged herpes simplex virus by Epstein–Barr virus (EBV)-transformed lymphoblastoid cell lines. *Virology*, **115**, 237–248 (1981).
55. R. Pollitt, F. Fenner and M. Davies, Sibs with mental and physical retardation and trichorrexis nodosa with abnormal amino acid composition of the hair. *Arch. Dis. Childh.*, **43**, 211–216 (1968).
56. A. Lehmann, C. Arlett, B. Broughton *et al.*, Trichothiodystrophy, a human DNA repair disorder with heterogeneity in the cellular response to ultraviolet light. *Cancer Res.*, **48**, 6090–6096 (1988).
57. K. Kraemer, J. DiGiovanna, A. Moshell, R. Tarone and G. Peck, Prevention of skin

cancer in xeroderma pigmentosum with the use of oral isotretinoin. *N. Engl. J. Med.*, **318**, 1633–1637 (1988).
58. B. Bridges, How important are somatic mutations and immune control in skin cancer? Reflections on xeroderma pigmentosum. *Carcinogenesis*, **2**, 471–472 (1981).
59. A. Kripke, Immunology of UV-induced skin cancer. *Photochem. Photobiol.*, **32**, 837–839 (1980).

13
Experimental Photocarcinogenesis

JAN C. VAN DER LEUN
University of Utrecht, Institute of Dermatology, The Netherlands

Right from the start of the problem of ozone depletion, the possibility of increasing skin cancer incidence has played a central role in the debate. I am not aware of any of the scientists involved who ever said that this was the most important consequence of ozone depletion, but it attracted most public attention. Increases in skin cancer have been on the front pages of newspapers many times; this has certainly helped to mobilize public attention and keep investigators in this area actively interested. This resulted in comparatively well developed quantitative predictions.

This chapter examines the relationship between ozone depletion and skin cancer incidence from the viewpoint of experimental photocarcinogenesis. Experimental work is necessary to collect the data required for good quantitative predictions.

The skin cancer problem posed by a depletion of ozone relates to man. Exposure to sunlight, even without ozone depletion, is considered to be the predominant cause of skin cancer in man. This applies most convincingly to squamous cell carcinomas; these occur almost exclusively on sun-exposed skin sites, and the incidence is clearly correlated with geographical latitude, being higher in the more sunny areas of the world. Basal cell carcinomas also have a preference for sun-exposed skin sites, but less exclusively so than squamous cell carcinomas; the incidence of basal cell carcinomas is also correlated with geographical latitude, but not as steeply as in the case of squamous cell carcinomas. This makes the case for a relationship between basal cell carcinoma

Ozone Depletion: Health and Environmental Consequences
Edited by R. Russell Jones and T. Wigley
© 1989 John Wiley & Sons Ltd

and sunlight less compelling. Yet, because of the analogies with squamous cell carcinoma, basal cell carcinoma is generally also considered to be a consequence of exposure to sunlight.

Frequently, basal cell carcinomas and squamous cell carcinomas are taken together as non-melanoma skin cancer. I will not deal with cutaneous melanomas, as these will be covered in Chapter 14. The non-melanoma skin cancers have a high incidence; in sunny areas with light-skinned people, they are by far the most common type of cancer. Medical treatment, if timely and given properly, is usually successful; if not treated properly, these cancers are dangerous and may ultimately kill the patient. In practice, the overall death rate is slightly less than 1%; the mortality is mainly due to the squamous cell carcinomas.

Quantitative estimates can be derived from the human data. The first quantitative prediction of the increase in the incidence of non-melanoma skin cancer from ozone depletion was based mainly on the correlation of incidence with geographical latitude, by the atmospheric physicist J.E. McDonald.[1] He predicted that ozone depletion would lead to a more-than-linear increase in skin cancer incidence, and coined the term 'amplification factor'. His amplification factor had a value of 6, indicating that a depletion in atmospheric ozone of 1% would ultimately lead to an increase in the incidence of non-melanoma skin cancer of 6%. McDonald was well aware that he had made many assumptions in his calculations. Looking backward, it is surprising that he came so close to the conclusions that are being reached now on the basis of many more data.

One of the assumptions required in this early prediction was that non-melanoma skin cancers were caused primarily by the UVB radiation (290–320 nm) in sunlight. That could not be concluded from epidemiological or clinical observations, as these deal with effects of exposure to full-spectrum sunlight. The fact that UVB radiation was the most effective wavelength range for the induction of skin cancer was already known, however, from early animal experiments by H. F. Blum.[2] This example typifies the problem. Epidemiological data are necessary but not sufficient for accurate predictions. More detailed information from animal experiments is also needed.

The animal experiments on photocarcinogenesis are usually done with mice, in recent years predominantly hairless albino mice.[3] The mice are exposed to ultraviolet radiation from lamps. The wavelengths may be varied by selecting the type of lamp and the filter, the dose by choosing the exposure time.

The mice are exposed regularly; for instance, daily. After a long period of such exposures (typically, several months to one year) tumours will develop, ultimately in all of the mice. The tumours formed are predominantly squamous cell carcinomas. The median tumour induction time (the time in which 50% of the animals develop tumours) is one of the measures for the effectiveness of the irradiation regimen chosen.

In this way it is possible to investigate the factors that influence the carcinogenic effectiveness of irradiations. One such factor, which has to be investigated carefully, is the way in which the carcinogenic effectiveness depends on the wavelength of the ultraviolet radiation; technically, this is called the action spectrum of ultraviolet carcinogenesis. This is needed because the change of irradiance caused by a depletion of ozone is spectrally very selective. Even in the narrow UVB range there is a lot of difference: the wavelengths immediately around 300 nm will increase much more steeply than wavelengths around 310 nm. So, if the effect depends more on 300 nm than on 310 nm, it will be much more sensitive to ozone depletion.

It is perhaps even more important to know the dose–effect relationship. If the animal receives more ultraviolet radiation, does it indeed develop more tumours and, if so, is that a steep relationship or is there perhaps a saturation?

DOSE–EFFECT RELATIONSHIP

Figure 1 shows a dose–effect relationship for photocarcinogenesis. It was determined in two series of experiments in which hairless albino mice Skh-hr 1 were exposed daily to the UVB radiation emitted by Westinghouse FS40 sunlamps. The figure gives the median tumour induction time as a function of the daily dose of UVB radiation administered. The horizontal scale encompasses a wide range of daily doses. At the upper end, the daily doses are only slightly below the doses inducing acute reactions in mouse skin. At the lower end, the doses are smaller by a factor of more than 30; these latter doses are appreciably

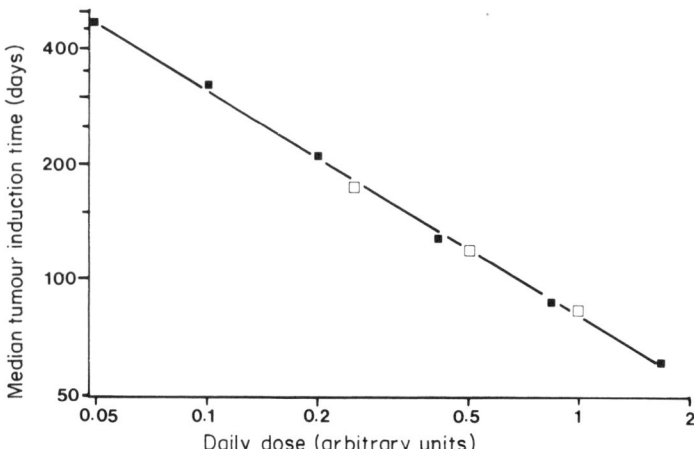

Figure 1 The relationship between the daily dose of UVB radiation and the time it takes for 50% of a group of mice to develop tumours. Both scales are logarithmic. Every dot represents an experiment with a group of hairless albino Skh-hr 1 mice. The daily doses are in arbitrary units; the highest daily dose applied was just below the dose causing acute reactions in the mouse skin. ■ from De Gruijl et al.[15]; □, from Slaper.[4]

smaller than the doses received in daily life by indoor workers in north-west Europe. Over this wide and realistic range of daily doses, the experimental dose-effect relationship shows no sign of deviating from a straight line. This implies that there is no indication of a saturation at the high doses and no indication, either, of a threshold dose; a daily dose so low that no carcinogenic effect would occur. The straight line in the log–log plot represents a power function.

For human populations, we are more interested in the incidence of skin cancer than in the time at which the cancers appear. The incidence is defined as the number of new patients per year per 100 000 of the population. An equivalent in the animal experiments is the tumour yield, the average number of tumours per mouse. The yield is determined by counting tumours rather than mice with tumours.

Slaper[4] found the yield for hairless albino mice, approximately, as

$$Y = SD^6 t^{10} \tag{1}$$

where Y is tumour yield, D the daily dose of ultraviolet radiation, t the time during which the daily exposures have been given, and, S a proportionality constant which includes the susceptibility of the strain of mice.

This is again a power relationship. The yield is steeply dependent on the daily dose of ultraviolet radiation administered, and even more steeply on time.

How do we translate these results to human populations? There are, of course, many differences between mouse and man, but the processes of photocarcinogenesis in the two species appear to be basically similar. The relationship between the skin cancer incidence in human populations and the ultraviolet radiation in their outdoor environment may be described by the same type of power relationship as that given in Equation 1, albeit with smaller exponents.[5] This speaks for a basic similarity, although the test is not as critical as one might wish, because of the large spread in the human data.

The description of the human data by power relationships appears to be the most useful for quantitative predictions; in this way we can utilize the detailed information given by the animal experiments, calibrated against the data collected in human populations. A large set of human data was carefully collected in the Third National Skin Cancer Survey in the USA.[6] From these data Slaper[4] derived the dependence of the incidence on dose; he found the exponent of dose to be 1.7 for basal cell carcinoma and 2.9 for squamous cell carcinoma.

This dependence of the incidence on the doses received regularly is less steep than for mice, but still more than linear. By the rules of calculus, these findings imply that, if the ultraviolet doses regularly received by a human population increase by 1%, the incidence of basal cell carcinomas will increase by 1.7% and that of squamous cell carcinomas by 2.9%. These numbers, 1.7 for basal cell carcinoma and 2.9 for squamous cell carcinoma, are called the 'biological amplification factors'.[7] These factors have nothing to do with the wavelengths responsible for photocarcinogenesis, but are entirely due to the curvature of the relationship between dose and incidence.

ACTION SPECTRUM

The dependence of the carcinogenic effectiveness of ultraviolet radiation on wavelength may be investigated only in experiments. Epidemiological observations offer hardly any information in this respect, and clinical observations none at all.

The experiments required are difficult and very time consuming. It is not easy to arrange monochromatic irradiations over objects as large as a cage with mice, daily, for a year or more, and for many wavelengths. Thus, until recently, there was no action spectrum for photocarcinogenesis. Investigators have produced several indications, however, that the action spectrum should be similar to that for ultraviolet erythema, which had been available for several decades.

In recent years, progress has been made in this area. Cole et al.[8] assembled the data on photocarcinogenesis collected over many years in the Skin and Cancer Hospital, Philadelphia, using hairless albino mice Skh-hr 1 exposed to radiations from different lamps with different filters. They tried to reconstruct one dose-effect relationship from all these data, by assuming hypothetical action spectra and computing from these effective doses for the various experiments. The best result was achieved with an action spectrum for the induction of oedema in mouse skin. This action spectrum was very similar to the action spectrum for ultraviolet erythema in human skin, thus confirming the idea that this was a good approximation of the action spectrum for photocarcinogenesis.

An even more direct programme to determine the action spectrum for photocarcinogenesis was carried out by Sterenborg[9] and Slaper[4] in a series of experiments also with Skh-hr 1 mice using eight different combinations of lamps and filters. Most of the spectra used overlapped to some extent, and the action spectrum was computed by deconvolution. Figure 2 shows the Sterenborg–Slaper action spectrum for photocarcinogenesis (solid line). The dashed line gives, for comparison, a recent standardized action spectrum for ultraviolet erythema. Comparison of the two curves confirms again that the approximation of the action spectrum for photocarcinogenesis by that for ultraviolet erythema was not bad. In the important wavelength range, the UVB, the correspondence is very close indeed. Deviations occur mainly in the UVC and UVA. From the peak near 300 nm towards the short-wavelength side, the carcinogenic effectiveness decreases. That is probably due to strong absorption of these short wavelengths in the horny layer. It indicates that there is not much reason to worry about a shift of the short wavelength cutoff of the solar spectrum, at least from the viewpoint of photocarcinogenesis. In any event, the shift will be small: for a 10% depletion of ozone the cutoff wavelength would shift by only 1 nm. The main reason for worry is the increase expected in the irradiance of UVB itself. Note that the

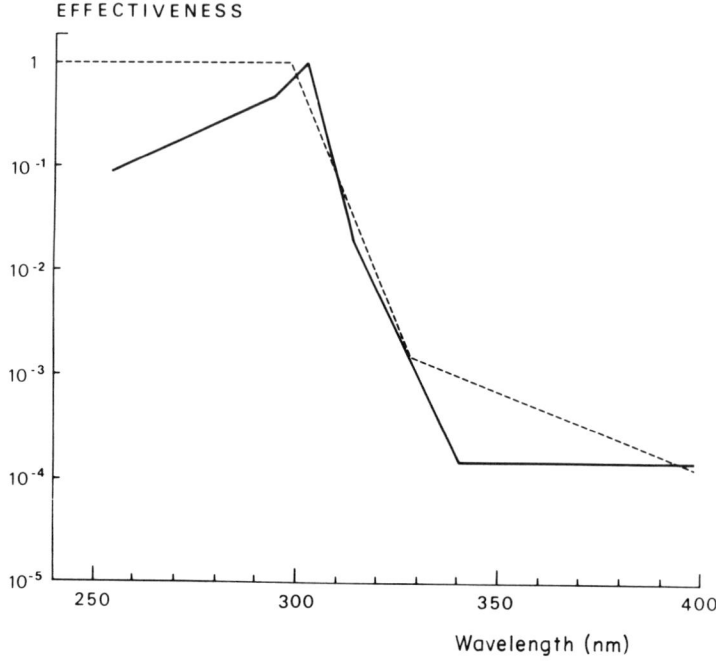

Figure 2 Two action spectra. The full line is an action spectrum for photocarcinogenesis in hairless albino mice, determined by Sterenborg[9] and Slaper.[4] The dashed line is a standardized action spectrum for ultraviolet erythema, adopted by the Commission Internationale de l'Éclairage.

carcinogenic effectiveness is on a log scale in Figure 2. The action spectrum shows that the effectiveness depends very steeply on wavelength in the wavelength range just above 300 nm, precisely where a steep increase of the solar irradiance is expected. That is the basis of another amplification, the 'optical amplification'.

The optical amplification factor (or radiation amplification factor) is defined as the percentage increase of the carcinogenically effective irradiance caused by a 1% decrease in total column ozone. It may be computed from the absorption spectrum of ozone and the action spectrum of the biological effect under consideration. As long as there was no action spectrum for photocarcinogenesis, it had to be approximated by assuming an action spectrum; that could be, for instance, the action spectrum for ultraviolet erythema or an action spectrum for DNA effects. The computation can now be made on the basis of a real action spectrum for photocarcinogenesis. A recent computation showed that, on the basis of the Sterenborg–Slaper action spectrum for photocarcinogenesis, the optical amplification factor has a value of 1.6 (Kelfkens and de Gruijl, personal communication). This is not far from the values calculated before, which usually ranged between 1.7 and 2.

We now proceed by combining the optical and biological amplifications found. A 1% decrease in ozone leads, according to the optical amplification factor, to an increase of the carcinogenically effective irradiance of 1.6%. According to the biological amplification factors, this leads to an increase in the incidence of basal cell carcinomas of $1.6 \times 1.7 = 2.7\%$; for the incidence of squamous cell carcinomas we find an increase of $1.6 \times 2.9 = 4.6\%$. When basal cell carcinomas and squamous cell carcinomas are put together as non-melanoma skin cancers, the increase in incidence will be about 3% for a 1% reduction of ozone.

These calculations provide the basis for calculating the effects of larger depletions. The depletion of total column ozone expected has varied over the years, with improvements in the atmospheric models. The investigators of biological effects cannot change their experiments and computations every year, so the expected biological changes are expressed as 'per percent of ozone depletion'.

IMMUNOLOGICAL CHANGES

Animal experiments have shown that ultraviolet radiation not only induces skin cancer, but also changes the immune response to the tumours induced. This was first observed in experiments in which tumours induced by ultraviolet radiation in a mouse were transplanted to genetically identical mice. The tumours were usually rejected, unless the receiving mouse had first been exposed to ultraviolet radiation.[10] This opened up an entire new field of investigation, called photoimmunology. It is not restricted to photocarcinogenesis; influences of ultraviolet radiation on the immune system also play a role in reactions of the skin to allergens[11] and to infections.[12]

The photoimmunological effects are caused mainly by UVB radiation, the spectral range that will increase in the event of ozone depletion. Several investigations have confirmed that the immunological changes induced by ultraviolet radiation also lead to an increased acceptance of primary tumours, induced by ultraviolet radiation in the animal itself. The effect is dose-related.[13]

These photoimmunological observations give us a deeper insight into the complicated processes leading to skin cancer.[14] In principle they do not require a change in the quantitative predictions of the increase in skin cancer incidence resulting from ozone depletion. These predictions were made mainly before the effects of ultraviolet radiation on the immune system were known, but they were entirely based on data from animal experiments and human epidemiology; in these situations, the effects of ultraviolet radiation on the immune system must have been present. These effects have, therefore, been implicitly taken into account.

In conclusion, the findings of photoimmunology do not give rise to additional worries from the viewpoint of photocarcinogenesis. It appears to be too early to give predictions on other potential consequences of the influence of ultraviolet radiation on the immune system. An ominous possibility is that increased UVB irradiation might lead to an increase in certain infectious diseases.

REFERENCES

1. J. E. McDonald, Relationship of skin cancer incidence to thickness of the ozone layer. *Congr.Rec.*, **117**, 3493 (1971).
2. H. F. Blum, *Carcinogenesis by Ultraviolet Light*. Princeton University Press, Princeton, NJ (1959).
3. P. D. Forbes, H. F. Blum and R. E. Davies, Photocarcinogenesis in hairless mice: dose response and the influence of dose-delivery. *Photochem. Photobiol.*, **34**, 361–365 (1981).
4. H. Slaper, *Skin Cancer and UV Exposure: Investigations on the Estimation of Risks*. PhD Thesis, University of Utrecht (1987).
5. T. R. Fears, J. Scotto and M. A. Schneiderman, Mathematical models of age and ultraviolet effects on the incidence of skin cancer in whites in the United States. *Am. J. Epidemiol.*, **105**, 420–427 (1977).
6. J. Scotto, T. R. Fears and J. F. Fraumeni, *Incidence of Non-melanoma Skin Cancer in the United States*. US Department of Health and Human Services, publ. NIH 82-2433 (1981).
7. J. C. van der Leun and F. Daniels Jr, Biologic effects of stratospheric ozone decrease: a critical review of assessments. In *Climatic Impact Assessment Program*, Monograph 5, Part 1, Chapter 7, pp 105–124, US Department of Transportation, Washington DC (1975).
8. C. A. Cole, P. D. Forbes and R. E. Davies, An action spectrum for photocarcinogenesis. *Photochem. Photobiol.*, **43**, 275–284 (1986).
9. H. J. C. M. Sterenborg, *Investigations on the Action Spectrum of Tumorigenesis by Ultraviolet Radiation*. PhD Thesis, University of Utrecht (1987).
10. M. L. Kripke, Photoimmunology: the first decade. *Curr. Probl. Dermatol.* **15**, 164–175 (1986).
11. F. P. Noonan, E. C. De Fabo and M. L. Kripke, Suppression of contact hypersensitivity in mice by UV radiation and its relationship to UV-induced suppression of tumor immunity. *Photochem. Photobiol.*, **34**, 683–689 (1981).
12. S. H. Giannini, Effects of UV-B on infectious disease. In T. G. Titus, ed., *Effects of Changes of Stratospheric Ozone and Global Climate*, UNEP/EPA Conference Proceedings, Vol. 2, pp 101–112 (1986).
13. J. C. van der Leun and F. R. de Gruijl, A systemic influence of UV radiation in photocarcinogenesis. In R.H. Douglas, J. Moan and F. Dall' Aqua, eds., *Light and Biology in Medicine*, Vol. 1, Plenum Press, New York, pp 293–299 (1988).
14. L. K. Roberts, W. E. Samlowski and R. A. Daynes, Immunological consequences of ultraviolet radiation exposure. *Photodermatology*, **3**, 284–297 (1986).
15. F. R. de Gruijl, J. B. van der Meer and J. C. van der Leun, Dose–time dependency of tumor formation by chronic UV-exposure. *Photochem. Photobiol.*, **37**, 53–62 (1983).

14
Epidemiology of Melanoma: Its Relationship to Ultraviolet Radiation and Ozone Depletion

J. MARK ELWOOD
University of Nottingham Medical School, UK

Melanoma is the most important type of skin cancer. In 1985 there were approximately 1000 deaths from melanoma in England and Wales, over twice as many as from all other skin cancers combined. For a population of 1 million people in the UK, there are about 45 new cases of melanoma, compared with some 450 registered non-melanoma skin cancers. Incidence data for non-melanoma skin cancer may suffer from substantial under-reporting. However, whereas non-melanoma skin cancer affects elderly people and has a high cure rate, melanoma affects a relatively young population and is a serious tumour with a substantial mortality; almost 40% of patients die from the disease within 5 years. The *deaths* per million are therefore about 20 per year for melanoma, compared with 10 per year for other skin cancers.

[Precise figures for 1985, the latest year available for England and Wales are 987 deaths from melanoma (455 men and 532 women), giving rates of 19 and 21 per million respectively. For other skin cancers there were 462 deaths (263 men and 199 women), giving rates of 11 and 8 per million respectively. Incidence data are available for 1984. There were 753 cases of melanoma in men and 1478 in women, giving rates of 31 and 58 per million. For other skin cancers there were 11 678 in men and 10 576 in women, giving rates of 482 and 414 per million respectively. The 5-year relative survival rates for melanoma patients registered from 1971 to 1973 were 48% for men and 69% for women. The crude survival rates were 41 and 62% respectively.[1-3]]

Ozone Depletion: Health and Environmental Consequences
Edited by R. Russell Jones and T. Wigley
© 1989 John Wiley & Sons Ltd

Melanoma is a malignant tumour that arises from melanocytes, the pigment-producing cells of the skin whose normal function is to respond to solar ultraviolet radiation by producing and disseminating pigment. As well as occurring in the skin, melanocytes are found internally, in the lining of the gut, the meninges of the brain, and the retina of the eye for example, and melanoma can also arise at these sites. This shows that the causes of melanoma are more complicated than simply ultraviolet radiation. However, 90% of melanomas arise from skin melanocytes, and it is this type of disease with which this chapter is concerned.

SKIN PIGMENTATION

The most striking thing about melanoma is that it is a disease of white-skinned people. Melanoma is much rarer in dark-skinned people, and when it does occur it has a different distribution and biological appearance than melanoma in white subjects. Even within white-skinned populations, melanoma is found more commonly in those with particularly light skin, which does not tan easily with solar exposure. Melanoma is also found much more commonly in light-skinned subjects who have many freckles, particularly in childhood, and in those who have more moles (benign acquired naevi), which are non-malignant collections of melanocytes that appear usually as small, flat, brown or black well demarcated areas on normal skin. These factors all point to a relationship between the malignant potential of melanocytes and features of their function and distribution in different people. These different skin types and pigmentation characteristics are most likely genetically determined characteristics over which we can exercise little control, although the number of naevi and freckles a subject has is also influenced by previous sun exposure.[4,5] The progression of a normal melanocyte to a malignant melanoma cell is a process that probably involves several stages, some of which may be genetically controlled, some of which may be influenced by ultraviolet exposure acting as a transforming agent producing a cancerous cell, and others which may be influenced by promoter or inhibitor actions of ultraviolet radiation acting either directly or through effects on the immune system and host defences. It is also possible that many factors other than sun exposure are involved in this process, although the evidence for the involvement of factors other than sunlight in melanoma in humans is very limited (Figure 1).

THE DISTRIBUTION OF MELANOMA IN POPULATIONS

Melanoma has not been produced and maintained in animal or *in vitro* models in ways that have allowed detailed work on the effects of ultraviolet radiation in its origins and development. This is in contrast to other types of skin cancer.

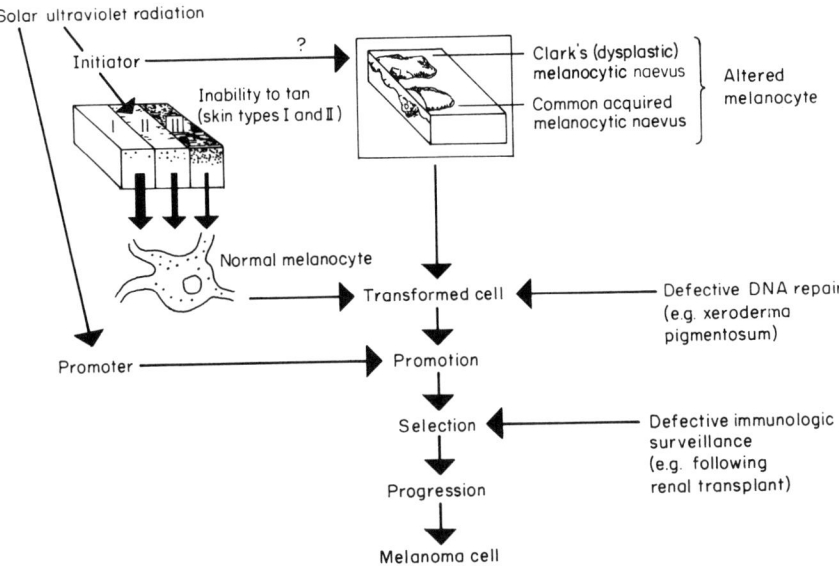

Figure 1 A diagram summarizing the various mechanisms by which ultraviolet radiation may influence the production of melanoma in the white population. (Reproduced by permission of Dr T.B. Fitzpatrick.)

Squamous cell cancers can readily be produced in several animal species by ultraviolet radiation, allowing the effects of ultraviolet radiation to be studied in great detail.[6] Experimental systems in fish and marsupials may in the future provide useful information, as discussed elsewhere in this volume. Thus in understanding the relationship between the occurrence of melanoma and solar or ultraviolet exposure we must go directly to the species with which we are concerned, and study the factors that influence the occurrence of melanoma in human populations using the methods of epidemiology.

GEOGRAPHY

One of the clearest observations is that if we look at large populations melanoma occurs much more commonly closer to the equator (Figure 2). This is seen very clearly in North America,[7] and is also seen, for example, within the UK, Norway, Australia and New Zealand.[8] The relationship is not clear across the whole of Europe, and this is because there are major and systematic differences in the pigmentation characteristics of people living in southern Europe as compared with northern Europe. This pattern of relationship with latitude I will discuss

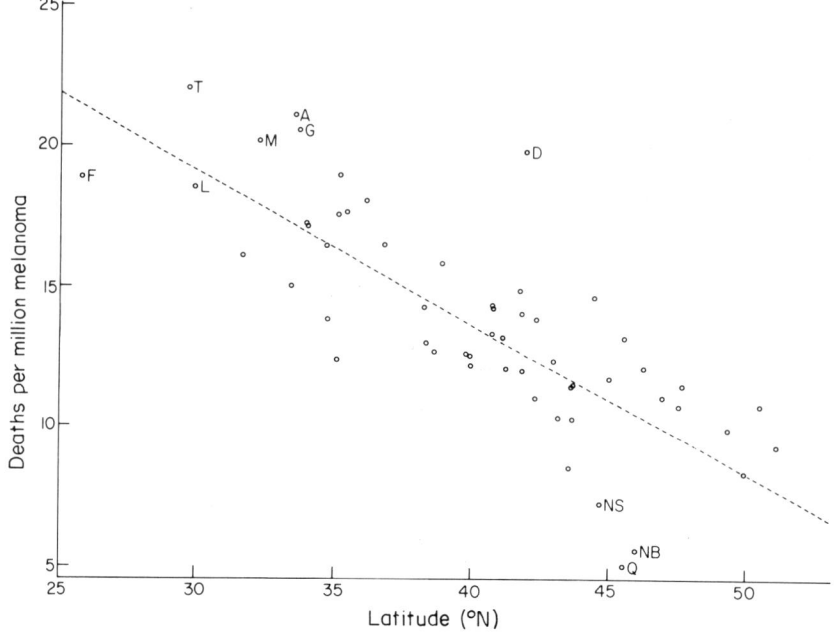

Figure 2   Relationship of age-standardized male mortality rates from melanoma for each US state and Canadian province, 1950–67, plotted against the latitude of the largest city. F, Florida; A, Alabama; NB, New Brunswick; D, Delaware; Q, Quebec; T, Texas; G, Georgia; L, Louisiana; M, Mississippi; NS, Nova Scotia. Newfoundland not shown (mortality 1.93, latitude 47.6°). (Reproduced from J.M. Elwood et al., Int. J. Epidemiol., 3, 325–332 (1974), with permission of Oxford University Press.)

later when we come to estimate the effects of changes in the ozone layer, and the gross pattern is very similar to the distribution for other sorts of skin cancer.

TIME TRENDS

The second major fact is that melanoma has been becoming much more common over the last few decades (Figure 3) in every white population on which we have information.[8-10] Unlike the geographical variation, this time trend is not seen for other types of skin cancer. At the same time, as the frequency has been increasing, the average age of patients with melanoma has fallen, and melanoma is now (in England and Wales) the fourth commonest cancer in women aged under 35, and the seventh commonest in men. Even in countries where the incidence has been much higher, such as Australia and New Zealand, this increase has also been seen, and in those areas melanoma is now the most common cancer in adults under age 40.[11]

Figure 3 Annual incidence rates per million population for cutaneous melanoma of all sites, by sex, England and Wales 1967–83. (New data, published by permission of the Office of Population, Censuses and Surveys.)

The time trend has been most marked for the leg in women and the back in men, and for the upper limb in both sexes, where the rates of increase have been considerably more than for the face. This is consistent with changes in exposure patterns due to changes in clothing and recreational habits.[8,10]

SOCIOECONOMIC DISTRIBUTION

Melanoma is more common in the more affluent; in the United Kingdom the mortality rate in social class 1 (managerial and professional groups) is about 70% higher than the rate in unskilled workers.[8]

SITE DISTRIBUTION

The distribution of melanoma on the body shows two major features. One is that differences between the sexes in the site distribution fit well with differential exposure by normal clothing patterns (Figure 4).[10] The second, perhaps even more important, is that if we calculate the frequency per unit area of skin, this is no higher on sites that are almost always exposed to the sun, such as the face, than it is on sites that are intermittently exposed, such as the back in men and the lower limb in women.[12] This is an important clue, which shows that the relationship of melanoma risk to solar exposure is not simple. The secular trend, socioeconomic and body site distributions, along with some other evidence, suggest that it is intermittent and irregular exposure of the body rather than continuous exposure that is important.

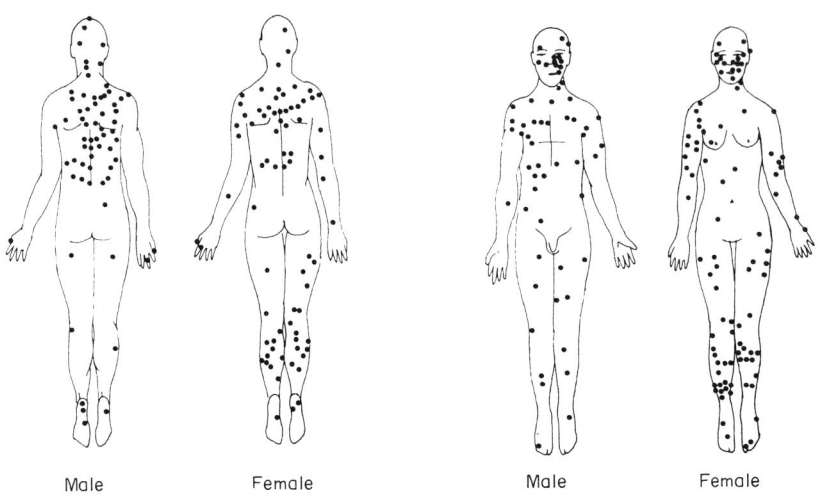

Figure 4 The body site distribution of a continuous series of 281 new cases of cutaneous melanoma of the superficial spreading or nodular type, seen in the A. Maxwell Evans Clinic in Vancouver from 1976 to 1979. (Reproduced from J.M. Elwood and R.P. Gallagher, *Can. Med. Assoc. J.*, **128**, 1400–1404 (1983[12] with permission.)

THE INTERMITTENT EXPOSURE HYPOTHESIS

This hypothesis, of an effect of intermittent rather than continuous solar exposure, has been tested in the largest study of melanoma in humans, carried out in Western Canada over the past decade.[13–17] In that study detailed information on lifestyle characteristics, including place of residence, occupation and holiday and recreational practices, was obtained from 665 subjects with newly diagnosed malignant melanoma, and an equal number of normal subjects of the same age and sex. The main results, applying to the 595 subjects with superficial spreading, nodular or unclassified melanoma, showed that the risk of melanoma increased with the amount of sun exposure which was obtained through holidays and recreational activities, and thus corresponded with intermittent exposure (Figure 5).[14] In regard to sun exposure achieved through regular occupation, those subjects with the greatest exposure, in other words those who had predominantly outdoor jobs, had a similar risk of melanoma to people who worked indoors all the time. The only increased risk was in subjects who had a small amount of occupational solar exposure, and this was a risk seen in people who had had seasonal employment or short periods of outdoor work, in other words exposure analogous to intermittent exposure.[14] This study is the clearest demonstration to date of the relationship between melanoma and individual sun exposure patterns.

A similar study was done in Western Australia, where of course the total level of sun exposure is much higher, and perhaps because of this did not show such a clear distinction between intermittent and continuous exposure. However, it showed that melanoma risk was increased with total sun exposure, with some very strong increases related, for example, to sunbathing and similar high-intensity exposures.[18–20] A large study in Denmark has recently been published whose results are very similar to those obtained by the Canadian study.[21,22] All these studies show that these associations with sun exposure are independent of the associations already mentioned with skin pigmentation and tendency to suntan. Thus, there are two separate sets of risk factors for melanoma. One is a set of factors related to pigmentation, naevi, freckles, and tendency to tan or to burn, all of which are skin characteristics related to genetic factors and perhaps to early exposure. The other is a set of factors involving exposure to sunlight, of which intermittent relatively intense exposure seems to be the important agent. One result of intense solar exposure of course is sunburn, and most studies also

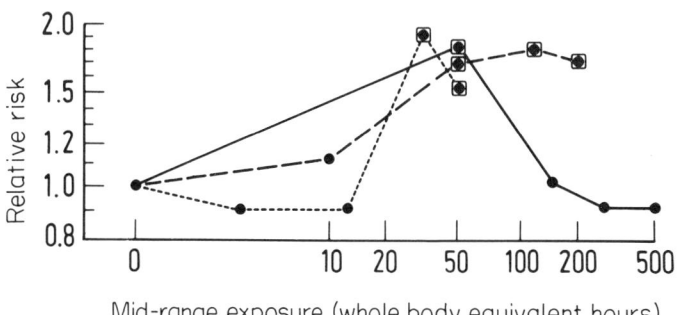

Figure 5 Summary of the results of the Western Canada Melanoma Study, showing the relative risk of melanoma in relationship to occupational, vacation and recreational sun exposures. Based on J.M. Elwood *et al.*, *Int. J. Cancer*, **35**, 427–433 (1985).- -, recreation; -----., vacation; —, occupation; □, significantly greater than 1.0.

show strong associations with histories of sunburn. It is not clear, however, whether the sunburn history is merely a clear indication of intermittent intense exposure, or whether on top of this there is a specific risk associated with the event of the sunburn.[15,18,23,24]

THE OZONE LAYER PROBLEM

I will now turn to discuss the problem of melanoma in relation to the ozone layer issue. My remarks so far should make it clear that the prediction of the effects of a decrease in atmospheric ozone on melanoma will not be simple. Whereas for other types of skin cancer we can assume rather simple models, such as the risk of skin cancer being proportional to total lifetime dose of solar ultraviolet radiation, we can make no such prediction for melanoma. The available evidence on melanoma in humans shows that the quantitative relationship between sun exposure and melanoma risk is complicated. The risk will depend not only on the dose of sun exposure received, but the regularity and timing of that dose, and the pigmentation characteristics of the person, which in turn are influenced by past exposure to the sun.

The highest melanoma risk occurs in individuals with low levels of chronic exposure but high levels of intermittent intense exposure; in fact, the typical exposure pattern of the affluent indoor worker in a developed country who can afford holiday or recreational activities involving considerable sun exposure.[14] Compared with such people, the risk in those who spend more time outdoors regularly, and so have a higher total dose of ultraviolet radiation, is lower. That presumably is due to the detrimental effects of ultraviolet being compensated by a protective effect of chronic exposure, which may be mediated through tanning, skin thickening or other mechanisms. Thus it is impossible to say for an individual what the effect of increasing the level of ambient ultraviolet radiation will be. For many, probably most, individuals, an increase in dose will move them towards the peak of increased risk, but for some an increase may decrease their risk by moving their total dose from peak levels towards the high levels that are related to protection.

If we move from consideration of the individual to the consideration of population groups, the issue becomes clearer. The easiest way to assess the effects of different levels of ultraviolet radiation exposure on a population is to compare populations who are similar in terms of genetic and pigmentation characteristics and lifestyle, and who live in places with differing amounts of exposure to ultraviolet radiation; that is, at different latitudes. We have already seen the regular trend in the incidence and mortality of melanoma, with higher rates closer to the equator.[7,8,10] There is no indication that any peak is reached; there is no point at which populations of similar pigmentation with higher exposures show a lower rather than a higher melanoma risk. Thus, the simplest way to estimate the effects of an increase in total ultraviolet radiation is to calculate a latitude equivalent. That is, we can measure the change in incidence or death rate from melanoma associated with a given difference in latitude, and obtain from estimates or measurements the difference in ultraviolet radiation related to that latitude change. Publications using this approach go back as far as 1974.

Estimates based on latitude and calculated ultraviolet levels

In 1974 my colleagues and I analysed death rates from melanoma in the continental USA and Canada.[7] We showed that the death rate varied dramatically with latitude. Between latitude 50° N (several Canadian provinces) to about 25° N (Florida and Texas), mortality rates more than doubled (Figure 2). The correlation coeffficient between melanoma mortality rate and latitude was very high; 0.78 for men and 0.72 for women. In other words, the melanoma death rate in any particular area can be predicted quite accurately from the latitude of that area without any other information. A variation in latitude of 2°, which is equivalent to 138 miles, carries with it a change in death rates from melanoma of about 10%.

In that study we worked with two colleagues, Drs Mo and Green of the Interdisciplinary Centre for Aeronomy and Other Atmospheric Sciences at the University of Florida, who had developed a calculation to estimate the annual ultraviolet flux at erythema-producing wavelengths from information on latitude and meteorological data on cloud cover. This calculated index of exposure was very strongly correlated with latitude (correlation coefficient of 0.89), so naturally melanoma mortality rates were strongly related to this index also. The data show that a 10% increase in received ultraviolet dosage would be expected to give an increase of 3.7–4.5% in the death rate from melanoma at latitude 50°, and 6.8–10.3% at latitude 30° (Table 1). These data showed somewhat higher values for men than women; for example, 4.4% in men compared with 3.0% in women at latitude 50° using the exponential model.

Two years later Fears, Scotto and Schneiderman of the National Cancer Institute performed a rather similar analysis for which they had data on the incidence as well as the mortality from melanoma, and related these measurements to latitude and to a calculated measure of ultraviolet radiation which, as with our measure, was strongly related to latitude.[25] Their data cover a slightly narrower range of latitude, and they calculated that a 10% increase in ultraviolet radiation would cause an increase in melanoma mortality of 7–12%, the higher figure applying to the more southerly latitudes, which already have higher rates. In simple terms, the idea of a 5–10% increase in mortality for a 10% increase in ultraviolet flux is reasonable.

Incidence rates vary more rapidly with latitude than do mortality rates, and therefore they predicted that a 10% increase in ultraviolet radiation would be likely to give a 14–24% increase in the incidence of melanoma. This is probably because in areas where melanoma is very common it is recognized earlier and

Table 1 Estimates of percentage increase in frequency of melanoma with a 10% increase in solar ultraviolet radiation

Ultraviolet data	Model	50° latitude		30° latitude		Notes	Reference
		Incidence	Mortality	Incidence	Mortality		
Calculation of erythema-weighted index	Linear		4.5		6.8	1	7
	Exponential		3.7		10.3		
Calculation of erythema-weighted index	Exponential	14.0	7.0	23.5	12.0	2	25
RB meter (1974)	Power	25.0		25.0		3	26
RB meter (1978–81)	Power						
	Trunk and lower limb						
	Crude	5.5		5.5		4	30
	Adjusted	3.5		3.5			
	Head, neck and upper limb						
	Crude	9.0		9.0			
	Adjusted	5.5		5.5			
	Total						
	Crude	6.7		6.7			
	Adjusted	4.2		4.2			
Calculated erythema-weighted estimate from NASA including satellite ozone column measurements	Power						
	Annual		3.2		3.2	5	31, 32
	Peak		7.0		7.0		
	Exponential						
	Annual		2.1		4.5		
	Peak		5.8		8.2		

All based on USA and Canada. Both sexes (simple average of sex specific results); white populations.
[1] Mortality data USA and Canada 1950–79 by state/province; 58 areas.
[2] Incidence data, Third National Cancer Survey (1969–71) nine areas; US mortality by state. Calculation based on latitude equivalent to change in ultraviolet.
[3] Incidence data, Third National Cancer Survey (1969–71) for four areas.
[4] Incidence data, Surveillance Epidemiology and End Results Program for seven areas. Crude results take account only of age; adjusted results are controlled for ethnic origin, hair or skin colour, suntan lotion use and hours spent outdoors. Total, for comparison is based on 67% trunk and lower limb and 33% head, neck and upper limb tumours.
[5] Mortality data by US county 1950–79. Estimates for changes in mean annual dose, and for change in peak doses (clear day in June). Estimates using DNA action spectrum were also made and were 1–8% higher than those shown.

treated earlier, and therefore a lower proportion of diagnosed patients die from the disease. Therefore those who live in a high-risk area in terms of incidence benefit from better survival, and so the difference in mortality is less marked.

Estimates using measured ultraviolet levels

These two studies used for data on ultraviolet radiation calculations based on distribution data and meteorological factors. In a further paper in 1977, Fears *et al.*, instead of using calculated indices of ultraviolet radiation, used measurements from Robertson–Berger (RB) meters, although the data were available only for four areas.[26] They used a power model, by which the calculated percentage changes are not dependent upon the initial latitude; the various mathematical models used in these studies are reviewed in the Appendix to this chapter. These calculations showed considerably higher effects, with an estimated 25% increase in incidence for a 10% increase in solar ultraviolet.

More recent work has been more sophisticated, although the overall results are not greatly different from the earlier work (Table 1). Fears and Scotto again used the measurements of ultraviolet radiation from RB meters placed by the National Weather Service at eight metropolitan areas in a 1983 paper, which deals with *non*-melanoma skin cancer.[27] They suggested that a 10% increase would be followed by increases from 18 to 30% in non-melanoma skin cancer rates. Estimates of the impact on malignant melanoma are given in a National Academy of Science (NAS) 1979 document[28] (page 100), suggesting percentage increases in melanoma incidence of 22% and in mortality of 14% for a 10% reduction in ozone; which other sources would suggest relates to a 20% increase in ultraviolet flux. The 1982 NAS report does not add anything beyond that given in 1979, merely commenting that the evidence for a direct relationship between ultraviolet and melanoma was actually weaker with the advent of more conflicting evidence.[29]

Scotto and Fears in 1987 used annual ultraviolet counts from RB meters in seven areas of the USA (Detroit, Seattle, Iowa, Utah, San Francisco, Atlanta and New Mexico), and data on melanoma from incidence registries (the SEER system).[30] They fitted a power model and presented analyses by sex and by body site of the melanoma divided into trunk and lower limb *versus* head, neck and upper limb. They obtained data on co-variates including ethnic origin, pigmentation characteristics, hours outdoors during weekdays and during weekends, and use of sunscreens, suntan lotion and protective clothing, from telephone interviews of at least 500 households in each area. However, the method uses data only for the general population, and does not involve more direct analysis, which would need such data on the melanoma patients also. Their results predict greater increases for females than for males, unlike the earlier work. The overall effects for a 10% increase in ultraviolet radiation are of a 5.5% increase for trunk and lower limb tumours and a 9% increase for head and upper limb tumours, averaged over the two sexes. Adjustment for the various co-variates reduces the predicted increases to a 3.5% increase for trunk and lower limb tumours, and 5.5% for head, neck and upper limb tumours. The adjustment for natural pigmentation and ethnic origin is appropriate, as previous models by failing to take this into account may be attributing to differences in ultraviolet radiation levels, differences in melanoma experience that are actually due to genetic or pigmentation characteristics. Whether one should adjust for behavioural characteristics such as time outdoors or use of sun protection is much more debatable. We are interested in the effect of a change in natural ultraviolet radiation levels on the human population, and if that human population changes its behaviour in response to the change in atmospheric conditions it would seem more logical to let such behavioural changes affect the calculated prediction, rather than calculate a theoretical predicted effect on the assumption that no behaviour changes will result. Adjustment for socioeconomic status might be useful but has not been done.

Further work has been done by Pitcher and partially reported, although it is still in progress.[31,32] He has been working with melanoma mortality data over a 30-year period, rather like the first papers in the series, and for ultraviolet measurements is using a calculation of ultraviolet flux based on NASA satellite data, including measurements of ozone concentrations in high atmospheric conditions. The models fitted are complex as they are fitted for the two sexes, for three different places covering a range of latitudes, and separately for changes in the annual ultraviolet flux and changes in the peak levels in clear summer conditions. He again finds larger effects for males than for females, and a larger effect when using the peak measurements than when using the annual measurements. His overall estimates of the percentage increase in melanoma mortality to be associated with 5% decrease in ozone level, on the assumption that this is roughly equivalent to a 10% in solar ultraviolet radiation, range from 2.1 to 7.0 at 50° latitude, and from 4.5 to 8.2 at 30° latitude.

The summary of the calculations made so far (Table 1) shows considerable variability, but the highest and now rather extreme estimates are those of Fears *et al.*, which were made using incidence data on just a few areas in the USA, and have produced estimates considerably higher than those given by any other method.[25,26] The studies using mortality data are all reasonably consistent, and would suggest that a 10% increase in solar ultraviolet radiation is related to an increase in melanoma mortality of somewhere between 2 and 10%.[7,25,29] This figure is not inconsistent with the most recent data from Scotto and Fears[30] on incidence, which has a range from 3.5 to 9% increase in incidence for a 10% increase in solar ultraviolet radiation depending on the models used and the adjustments made.

Time trends in melanoma in relation to changes in ultraviolet radiation

If melanoma rates were otherwise stable over time, this predicted increase would give us the future trend. However, melanoma rates are far from stable, and have been increasing rapidly over recent decades. Can we say that the trend already seen is due to ozone depletion and a consequent increase in ultraviolet radiation?

Some data are available from England, where measurements of total ozone amounts for Oxford from 1950 to 1972 showed a very steady 10% *increase* over this period.[33] That is, there was an increase in atmospheric ozone, and the authors calculate a corresponding *decrease* in ultraviolet radiation levels. (At least until 1972; their data for the years 1972–76 show an increase in ultraviolet radiation,

EPIDEMIOLOGY OF MELANOMA   181

which they regard as due to the particularly good summers of 1975 and 1976.) During this time the incidence and mortality rates of melanoma in Oxford and elsewhere in the UK showed a steady rise.

These authors also compared ultraviolet flux with incidence and mortality rates from other skin cancer, but their logic is curious. Other skin cancer rates showed a decrease from 1951 to 1961, in parallel with the decrease in ultraviolet, and therefore they regard that as evidence supporting a direct relationship. All these rates showed an abrupt change with an increase since 1961, which of course is not in keeping with the ultraviolet change (and is so abrupt that it suggests a recording artefact). They failed to take into account any latency period, and in their discussion comment that 'irrespective of the length of time for a carcinoma to develop ... a change in radiation in a year must produce a proportional change in incidence in that year'.[33] This is illogical.

More detailed work from the USA has recently been reported on the results from RB meters at eight centres in the USA, ranging from 30° to 37° latitude, from 1974 to 1985.[34] For all stations, there was a slight *downward* trend in solar ultraviolet radiation. The estimated average annual change varied from −1.1% in Minneapolis to −0.4% in Philadelphia; the overall change was a decrease of 0.7% per year from 1974 to 1975. The authors point out that these results are consistent with earlier reports from RB meters, and from studies using Dobson ozone measurements between 1970 and 1982 (which presumably showed an increase). They point out that the RB meter is relatively insensitive to the shorter UVB wavelengths in comparison with the erythema action spectrum, and therefore a change in the RB meter reading produced by changes in ozone will be likely to be less than the change in the erythemal dose. They suggest that the lack of any increase in UVB shows that the role of physical and meteorological factors is greater than expected and may overcome the effects of predicted changes in stratospheric ozone.

The inconsistency between the secular changes seen in the ultraviolet data[34] and the evidence of a small but clear *decrease* in stratospheric ozone over the Northern Hemisphere (discussed elsewhere in this volume) is curious. It has been suggested that, as the RB meters are located at airports, the downward trend seen in surface ultraviolet radiation might be due to local conditions, such as increasing atmospheric pollution. The relevant question here is whether past changes in melanoma rates are related to changes in ground level ultraviolet radiation. If so, any effects of local conditions are relevant. The issue that remains is whether the results from this radiation monitoring system apply to surface ultraviolet fluxes in the major urban areas where most people live. The answer is uncertain, but there is certainly no evidence of a major *increase* in ultraviolet levels.

Thus, the available data suggest that the large increase in melanoma that has been observed cannot be explained by an increase in ground level ultraviolet radiation. If the trends are the effects of ultraviolet radiation, they must be caused by an increase in personal exposure due to differences in activity and clothing habits and, as pointed out above, several facets of the increase, particularly the site specificity, appear to conform to this hypothesis.

Prediction of future trends in melanoma

This makes the prediction of future increases in melanoma very difficult. The changes in melanoma rate observed over the last decades have been very large and very rapid. If they are due to behavioural changes, these behavioural changes must have had dramatic effects on the amount of ultraviolet radiation received, and its influence on melanoma. Measurements made using personal dosimeters have suggested that a holiday involving sunbathing at a southern European location can double the total amount of solar ultraviolet received by a predominantly indoor-living individual in a northern European country; such results come from the UK, the Netherlands, Norway and Sweden.[37-41] This doubling of ultraviolet received by the body is of course a much bigger effect than most predictions of ozone layer effects. The question remains, however, as to whether behavioural changes provide a totally adequate explanation for the change in melanoma risk, or whether other factors are involved. The difficulty is the lack of any coherent direct evidence of the influence of factors other than solar exposure on melanoma. Tantalizing clues suggesting a role for hormonal factors, dietary factors, and so on, can be obtained, but there is little consistency.

The 'worst case' scenario for a prediction of melanoma risk in the future might be that the increases seen over the last few decades will continue, and to these we would add the predicted effects of an increase in ultraviolet due to ozone layer effects. The result is a suggestion of a further rapid increase in melanoma that will make it before long one of the most common adult cancers. However, the added effect of ozone layer changes in the equation is fairly small. Large further increases in melanoma incidence are predicted, with or without inclusion of an ozone depletion factor. Such a prediction was made some 10 years ago by Dr Lee in the USA and has been confirmed clearly by the actual changes seen.[42] Furthermore, the concept that melanoma could become one of the more common adult cancers is again not unrealistic; in Australia and New Zealand, malignant melanoma is now the most common invasive tumour in adults between the ages of 20 and 40.[11]

The evidence suggests that the factors that may be able to prevent that further increase are changes in behaviour; in our attitudes to sun exposure, our habits of sun exposure, and our use of protective practices and materials. Efforts at public education have been pursued most vigorously in Queensland, Australia, which has the highest melanoma rate in the world, and there is evidence over recent decades that the death rate from melanoma has stabilized.[43] This, however, appears to be mainly due to improvements in early diagnosis and therapy, and the incidence rate of melanoma in Queensland appears to be still rising despite its already very high level, although the available data are several years old.[44] More recent data from Queensland are urgently required to show whether public education efforts there have in fact managed to stem the increasing trend in melanoma incidence.

Latency of effect

A complex issue that relates to all calculations of trend is the time relationship between exposure to ultraviolet radiation and the occurrence of melanoma.

Useful evidence on this comes from studies of migrants who move from less sunny to more sunny countries; European migrants to Australia, to New Zealand, and Central and Northern European Jewish migrants to Israel.[19,45,46] The Australian studies show that people who migrate to Australia before the age of 10 have similar rates of melanoma to those who were born in Australia, but subjects who migrate at ages 15–19 or older have much lower rates.[19] So high rates relate to residence in the sunnier countries at later than age 5–9 but before 15–19, suggesting that the critical age for exposure may be between 10 and 19 years. It may be relevant that the strongest pigmentary predictor of risk of melanoma in an individual is their number of benign acquired naevi.[4,5] These normally develop at puberty, and increase in frequency up to about age 20, when the number stabilizes, and in later life may decrease. Thus this maximum sensitivity to sun exposure in the teenage years and early adult life is related to the time of the maximum frequency of normal benign naevi.[4,5]

Some other evidence suggests that even earlier exposures might be important. Several studies have suggested that sunburn, and particularly sunburn in childhood, is important. In some of these the investigators asked only about sunburn in childhood, and therefore could not differentiate the effects of sunburn at different ages. In a recent study in the UK, we have asked the same question about sunburn at different periods of life, and the data suggest that sunburn in childhood has a particularly strong effect (unpublished data). However, all such studies are based on retrospective histories given by individuals when asked questions concerning several decades of life. We must be aware that the factor that comes out as strongest may be the factor that is biologically the major risk factor, but may also be the question that is most easily answered and therefore gives clearest responses. Sunburn in childhood may be a more dramatic or easily remembered event than sunburn in adult life, and may therefore be picked up more easily by the studies.

The concept of the importance of ultraviolet radiation in early life is consistent with the effects seen with many other cancer-causing agents, such as ionizing radiation, where for example subjects exposed through the atomic bomb attack had a high risk of cancer if they were younger at the time of exposure. Such effects are typical of cancer initiators; that is, agents producing the first stage in the transformation from a normal cancerous cell. In its production of non-melanoma skin cancer, ultraviolet radiation acts as a classic initiator agent, as detailed and quantitative studies in animals have shown. However, the same argument may not hold for melanoma. Melanomas have not been produced experimentally by ultraviolet radiation acting as an initiator in animals. The experimental work on melanoma shows that ultraviolet radiation can act as a tumour promoter on cells that have already been transformed by a chemical carcinogen, and also that ultraviolet radiation can have immunosuppressive effects. In melanoma, the initiator action may be a genetic defect or a defect closely related to the aetiology of the apparently benign acquired naevus, and ultraviolet radiation may be acting as a promoter.[6]

The concept of a promoter action of ultraviolet radiation raises the question of a short-term effect, as the action of a promoter could logically be followed by an increase in observed cancer incidence a few months or years later. The possibility of a short-term effect in melanoma was suggested by data from the USA, suggesting that melanoma frequency was increased approximately 2 years after years of peak sunspot activity.[36] Attempts to verify this observation in other sets of data have given inconsistent results, and there is a difficulty in distinguishing a real correlation from spurious associations. However, recent work on short-term fluctuations in measured ultraviolet levels in Philadelphia shows that during years of maximum sunspot activity ultraviolet radiation at ground level is decreased rather than increased.[35]

In discussion at this meeting it was said that the increase in solar radiation at high sunspot periods is limited to wavelengths below 240 nm, while stratospheric ozone levels are increased by about 2%, resulting in a decrease in ultraviolet intensity at ground level (see p. 193).

Other evidence on short-term fluctuations is the work by Swerdlow in the UK, correlating melanoma peaks with years of particularly sunny summers, such as 1975 and 1976.[47] This is not open to the same misinterpretation in terms of the effect of sunspots, but more data are needed to confirm the correlation. Several studies show a higher frequency of melanoma newly diagnosed in the summer rather than in the winter months, but it is difficult to disentangle a real short-term effect of summer sunshine on incidence from an effect on diagnosis, as people may be much more ready to recognize and do something about a pigmented lesion in the summer months. More persuasive as evidence of a short-term effect of sunlight is the work on the pathological appearances of excised naevi, showing evidence of more histological mitotic activity and dedifferentiation in those excised during and towards the end of the summer.[48]

All these issues of course make prediction of major trends very difficult, and the evidence is strongly suggestive that some or all of the ultraviolet effect has a long-term basis; thus the sun behaviour and sun exposure of teenagers and young adults today will determine their melanoma risk over the next 20–40 years. Thus the beneficial effects of changes in behaviour, and the detrimental effect of changes in the ozone layer, will be seen in future decades rather than instantaneously.

CONCLUSIONS

Some points are worthy of emphasis. The evidence shows that melanoma is related to solar and thus, it is assumed, to ultraviolet radiation exposure; but the relationship of melanoma risk to quantity of exposure is not linear, and the time relationship is uncertain. Melanoma has a different relationship to solar exposure than do the other types of skin cancer.

The major determinants of the risk of a melanoma in an individual subject form three sets:

(1) Pigmentation factors, skin reaction to sun exposure, freckles and naevi.
(2) Personal behaviour in regard to sun exposure.
(3) The ultraviolet radiation levels at the places on the Earth's surface where the subject lives and travels to.

The response to the melanoma problem should take account of these three sets of factors, leading to efforts to:

(1) Identify, monitor and use active preventive measures on high risk subjects.
(2) Change personal behaviour.
(3) Improve or at least minimize deterioration in the environment.

Control of the ozone layer problem is, in regard to melanoma, an important issue and success will aid other approaches to the control of this disease.

REFERENCES

1. *OPCS Mortality Statistics, Cause 1985.* Series DH2 No. 12, HMSO, London (1987).
2. *OPCS Cancer Statistics, Registrations 1984.* Series MB1, No. 16, HMSO, London (1988).
3. *OPCS Cancer Statistics, Survival 1971–73.* Series MB1, No. 3, HMSO, London (1980).
4. B. K. Armstrong and D. R. English, The epidemiology of acquired melanocytic naevi and their relationship to malignant melanoma. In J. M. Elwood, ed., *Melanoma and Naevi, Pigment Cell,* Vol. 9, Karger, Basle, pp 27–47 (1988).
5. B. K. Armstrong, N. de Klerk and C. Holman, Etiology of common acquired melanocytic naevi: constitutional variables, sun exposure and diet. *J. Natl. Cancer Inst.,* 77, 329–335 (1986).
6. J. M. Elwood, Initiation and promotion actions of ultraviolet radiation on malignant melanoma. In M. Borzsonyi, N. E. Day, K. Lapis and H. Yamasaki, eds, *Models, Mechanisms and Etiology of Tumor Promotion,* IARC Scientific Publication No. 56, MRC-OUP, Lyon, pp 421–525 (1984).

7. J. M. Elwood, J. A. H. Lee, S. D. Walter, T. Mo and A. E. S. Green, Relationship of melanoma and other skin cancer mortality to latitude and ultraviolet radiation in the United States and Canada. *Int. J. Epidemiol.*, **3**, 325–332 (1974).
8. J. A. H. Lee, Melanoma and exposure to sunlight. *Epidemiol. Rev.*, **4**, 110–136 (1982).
9. O. M. Jensen and A. M. Bolander, Trends in malignant melanoma of the skin. *World Health Stat. Q.*, **33**, 2–26 (1980).
10. J. M. Elwood and J. A. H. Lee, Recent data on the epidemiology of malignant melanoma. *Sem. Oncol.*, **2**, 149–154 (1975).
11. J. M. Elwood, K. R. Cooke, B. D. Coombs, B. Cox, J. E. Hand and D. C. G. Skegg, A strategy for the control of malignant melanoma in New Zealand. *N. Z. Med. J.*, **101**, 602–604 (1988).
12. J. M. Elwood and R. P. Gallagher, Site distribution of malignant melanoma. *Can. Med. Assoc. J.*, **128**, 1400–1404 (1983).
13. J. M. Elwood, R. P. Gallagher, G. B. Hill, J. J. Spinelli, J. C. G. Pearson and W. Threlfall, Pigmentation and skin reaction to sun as risk factors for cutaneous melanoma; the Western Canada Melanoma Study. *Br. Med. J.*, **288**, 99–102 (1984).
14. J. M. Elwood, R. P. Gallagher, G. B. Hill and J. C. G. Pearson, Cutaneous melanoma in relation to intermittent and constant sun exposure: the Western Canada Melanoma Study. *Int. J. Cancer*, **35**, 427–433 (1985).
15. J. M. Elwood, R. P. Gallagher, J. Davison and G. B. Hill, Sunburn, suntan and the risk of cutaneous malignant melanoma: the Western Canada Melanoma Study. *Br. J. Cancer*, **51**, 543–549 (1985).
16. J. M. Elwood, R. P. Gallagher, A. J. Worth, W. S. Wood and J. C. G. Pearson, Etiological differences between subtypes of cutaneous malignant melanoma: Western Canada Melanoma Study. *J. Natl. Cancer Inst.*, **78** 37–44 (1987).
17. R. P. Gallagher, J. M. Elwood, W. J. Threlfall, J. J. Spinelli, S. Finsham and G. B. Hill, Socio-economic status, sunlight exposure, and risk of malignant melanoma: the Western Canada Melanoma Study. *J. Natl. Cancer Inst.*, **79**, 647–652 (1987).
18. C. D. J. Holman and B. K. Armstrong, Pigmentary traits, ethnic origin, benign nevi, and family history as risk factors for cutaneous malignant melanoma. *J. Natl. Cancer Inst.*, **72**, 257–266 (1984).
19. C. D. J. Holman and B. K. Armstrong, Cutaneous malignant melanoma and indicators of total accumulated exposure to the sun: an analysis separating histogenetic types. *J. Natl. Cancer Inst.*, **73**, 75–82 (1984).
20. C. D. J. Holman, B. K. Armstrong and P. J. Heenan, Relationship of cutaneous malignant melanoma to individual sunlight exposure habits. *J. Natl. Cancer Inst.*, **76**, 403–414 (1986).
21. A. Osterlind, M. A. Tucker, K. Hou-Jensen, B. J. Stone, G. Engholm and O. M. Jensen, The Danish case–control study of cutaneous malignant melanoma. I. Importance of host factors. *Int. J. Cancer*, **42**, 200–206 (1988).
22. A. Osterlind, M. A. Tucker, B. J. Stone and O. M. Jensen, The Danish case–control study of cutaneous malignant melanoma. II. Importance of UV-light exposure. *Int. J. Cancer*, **42**, 319–324 (1988).
23. A. Green, Y. Siskind, C. Bain and J. Alexander, Sunburn and malignant melanoma. *Br. J. Cancer*, **51**, 393–397 (1985).
24. R. M. MacKie and T. G. Aitchison, Severe sunburn and subsequent risk of primary cutaneous malignant melanoma in Scotland. *Br. J. Cancer*, **46**, 955–960 (1982).
25. T. R. Fears, J. Scotto and M. A. Schneiderman, Skin cancer, melanoma and sunlight. *Am. J. Public Health*, **66**, 461–464 (1976).
26. T. R. Fears, J. Scotto and M. H. Schneiderman, Mathematical models of age and

ultraviolet effects on the incidence of skin cancer among whites in the United States. *Am. J. Epidemiol.*, **105**, 420–427 (1977).
27. T. R. Fears and J. Scotto, Estimating increases in skin cancer morbidity due to increases in ultraviolet radiation exposure. *Cancer Invest.*, **1**, 119–126 (1983).
28. NAS, *Protection against Depletion of Stratospheric Ozone by Chlorofluorocarbons.* National Academy of Sciences, Washington DC (1979).
29. NAS, *Causes and Effects of Stratospheric Ozone Reduction: An Update.* National Academy of Sciences, Washington DC (1982).
30. J. Scotto and T. R. Fears, The association of solar ultraviolet radiation and skin melanoma incidence among Caucasians in the United States. *Cancer Invest.*, **5**, 275–283 (1987).
31. H. M. Pitcher, Examination of the empirical relationship between melanoma death rates in the United States 1950–1979 and satellite-based estimates of exposure to ultraviolet radiation. In press.
32. J. S. Hoffman and J. D. Longstreth, Malignant melanoma. In J. S. Hoffman, ed., *An assessment of the risks of stratospheric modification*, Vol. III. US Environmental Protection Agency, Washington DC (1986).
33. J. F. Leach, P. C. Beadle and A. R. Pingstone, Effect of ozone variation on disease in Great Britain. 1. Skin cancer. *Aviat. Space Environ. Med.*, **49**, 512–516 (1978).
34. J. Scotto, G. Cotton, F. Urbach, D. Berger and T. Fears, Biologically effective ultraviolet radiation: surface measurements in the United States, 1974 to 1985. *Science*, **239**, 762–764 (1988).
35. D. S. Berger, Fluctuations and trends in environmental UV loads. In W. F. Passchier and B. F. M. Bosnjakovic, eds, *Human Exposure to Ultraviolet Radiation: Risks and Regulations*. Elsevier, Amsterdam, pp 213–221 (1987).
36. A. Houghton, E. W. Munster and M. V. Viola, Increased incidence of malignant melanoma after peaks of sunspot activity. *Lancet* **i**, 759–760 (1978).
37. J. H. Bernhardt and R. Matthes, Quantitative description of exposure to UV and methods of measurements. In W. F. Passchier, and B. F. M. Bosnjakovic, eds, *Human Exposure to Ultraviolet Radiation: Risks and Regulations*. Elsevier, Amsterdam, pp 201–212 (1987).
38. B. L. Diffey, O. Larko and G. Swanbeck, UV-B doses received during different outdoor activities and UV-B treatment of psoriasis. *Br. J. Dermatol.*, **106**, 33–41 (1982).
39. H. Slaper and J. C. van der Leun, Quantitative modelling of skin cancer incidence. In W. F. Passchier and B. F. M. Bosnjakovic, eds, *Human Exposure to Ultraviolet Radiation: Risks and Regulations*. Elsevier, Amsterdam, pp 155–171 (1987).
40. A. Schothorst, H. Slaper, D. Telgt, B. Alhadi and D. Suurmond, In W. F. Passchier and B. F. M. Bosnjakovic, eds, *Human Exposure to Ultraviolet Radiation: Risks and Regulations*. Elsevier, Amsterdam, pp 269–273 (1987).
41. E. Kivisakk, Intentional exposure to ultraviolet radiation: risk reduction and present regulations. In W. F. Passchier and B. F. M. Bosnjakovic, eds, *Human Exposure to Ultraviolet Radiation: Risks and Regulations*. Elsevier, Amsterdam, pp 443–454 (1987).
42. J. A. H. Lee, *An Update on the Epidemiology of Malignant Melanoma*. Report to the Environmental Protection Agency (1985).
43. G. R. McLeod, Control of melanoma in high-risk populations. In *Pigment Cell*, Vol. 9. Karger, Basel, pp 131–139 (1988).
44. B. Armstrong, C. Holman, J. Ford and T. Woodings, Trends in melanoma incidence and mortality in Australia. In Magnus, ed., *Trends in Cancer Incidence. Causes and Practical Implications*. Hemisphere, Washington, 339–417 (1982).

45. K. R. Cooke and J. Fraser, Migration and death from malignant melanoma. *Int. J. Cancer*, **36**, 175–178 (1985).
46. D. Anaise, R. Steinitz and N. Ben Hur, Solar radiation: a possible etiological factor in malignant melanoma in Israel: a retrospective study (1960–72). *Cancer*, **42**, 299–304 (1978).
47. A. J. Swerdlow, Incidence of malignant melanoma of the skin in England and Wales and its relationship to sunshine. *Br. Med. J.*, **ii**, 1324–1327 (1979).
48. B. Armstrong, P. Heenan, V. Caruso, R. Glancy and C. Holman, Seasonal variation in the junctional component of pigmented naevi. *Int. J. Cancer*, **34**, 441–442 (1984).

APPENDIX: MATHEMATICAL MODELS

There have been three models used to relate changes in ultraviolet or ozone levels to changes in melanoma incidence or mortality: linear, exponential and power models. They are described below. The basic model is given, and also the formula by which a proportional change in melanoma frequency is related to a proportional change in the predictor variable, whether that is the ultraviolet level or ozone level. In both the linear and the exponential models, where the baseline ultraviolet radiation level is higher, the percentage change in melanoma frequency produced by a given percentage change in ultraviolet radiation will be greater. On the power model, the percentage change produced is independent of the baseline level of ultraviolet radiation, and therefore independent of latitude.

Let

M = incidence or mortality rate from melanoma
M_0 = rate at baseline ultraviolet level
M_1 = rate at higher ultraviolet level
V = level of ultraviolet (e.g. annual erythemal dose)
V_0 = baseline level
b = regression coefficient of M on V
a = intercept in regression equation

The objective is to calculate the proportional change in M for a given increase in V, that is

$$\frac{M_1 - M_0}{M_0} = \frac{M_1}{M_0} - 1$$

Formulae are shown for a 0.1 (10%) increase in V; for any other proportional increase in V, say p, replace 0.1 in the following by p and replace 1.1 by $(1+p)$.

Linear model
Used in Elwood *et al.*[7]

$$M = bV + a$$
$$M_0 = bV_0 + a$$
$$M_1 = 1.1bV_0 + a$$

thus
$$M_1 - M_0 = 0.1bV_0$$
proportional change in $M_0 =$
$$\frac{M_1 - M_0}{M_0} = \frac{0.1bV_0}{bV_0 + a}$$
$$= \frac{0.1}{1 + a/bV_0}$$
$$= \frac{0.1}{1 + (a/b)(1/V_0)}$$

Thus, to calculate the proportional change, we need a, b and V_0; the proportional change increases as V_0 increases.

Exponential model
Used in Elwood et al.,[7] Fears et al.[25] and Pitcher.[32]
$$\ln M = bV + a$$
$$\ln M_0 = bV_0 + a$$
$$\ln M_1 = 1.1bV_0 + a$$
$$\ln(M_1/M_0) = \ln M_1 - \ln M_0$$
$$= 0.1bV_0$$
$$(M_1/M_0) - 1 = \exp(0.1bV_0) - 1$$

Thus, to calculate the proportional change, we need b and V_0; the proportional change increases as V_0 increases.

Power model
Used by Fear et al.,[26] Scotto and Fears[30] and Pitcher.[32]
$$\ln M = b \ln V + a$$
$$\ln M_0 = b \ln V_0 + a$$
$$\ln M_1 = b \ln(1.1V_0) + a$$
$$= b(\ln 1.1 + \ln V_0) + a$$
$$\ln M_1 - \ln M_0 = b \ln 1.1 = \ln(M_1/M_0)$$
$$(M_1/M_0) - 1 = \exp(b \ln 1.1) - 1$$

Thus, to calculate the proportional change, we need only b; the proportional change is independent of V_0.

Discussion Period 4

Professor Scorer (Imperial College) I am very much concerned with air pollution. Would you consider that the reduction in air pollution has given people more regular sunshine so that they are less likely to get melanoma, or do you think it has increased melanoma risk by increasing the amount of ultraviolet radiation?

Rona Mackie I can answer that to a limited extent from my Glasgow study, in which we have measured ultraviolet at ground level for many years and found an increase correlating with smokeless fuel introduction. I think there is no doubt that more ultraviolet reaches UK cities now. But of course the behaviour of the individuals in UK cities has changed rather more, I think, than the ultraviolet radiation reaching them. I do not think we can explain metropolitan melanoma on changes in smokeless fuel.

Dr Morton There has been a considerable increase in the use of suntanning lamps, and from the last speaker's talk it is hard to guess what the effect will be; on the one hand the user is increasing exposure to ultraviolet, but, on the other hand, if used regularly these people have a constant tan which may be protective. One also has to consider wavelength. These lamps are predominantly UVA, although I believe they do give out some UVB, or some may do so. Is there an effect of wavelength on the induction of melanoma?

Dr van der Leun The use of artificial sources is indeed widespread and the recent types are based on longer wavelengths, mainly UVA, and less UVB. What effect this will have on skin cancer incidence is not completely clear. The action spectrum I showed indicates that for the same degree of pigmentation UVA is about as carcinogenic as UVB; that is, for squamous cell carcinoma. There is indeed some protection from tanning but not much, and certainly not from UVA-induced tan. That has a very limited protective action.

With regard to your question about the action spectrum for melanoma, that simply is not known. It is generally assumed to be similar to the action spectra of other skin cancers, but there are no data from animal models. As regards the other skin cancers the relevant wavelengths are known only from experimental work. Hopefully, there will soon be experimental data on animals that develop melanoma after exposure to ultraviolet light.

Dr Arlett One of the things that needs to be addressed is the interaction between the different wavelengths. Several cellular studies have provided useful information, and it has been shown that in some situations predosing with UVA actually protects and in other cases predosing with UVA actually sensitizes cells. So this is a 'magic garden', which has not yet been explored. If we assume that changes in the ozone layer have a differential effect on UVA and UVB, such effects will need to be taken into account.

Dr Thomas Barnsley How accurate are the data on the incidence of melanoma and other skin cancers? It must depend on reporting these cases to the cancer registry. I know for a fact that not all dermatologists and certainly many general practitioners will not be reporting these at all.

Professor Elwood You are right. Incidence data are probably underestimated and the degree of error varies in different parts of the world. The better registries obtain their data directly from the pathology laboratories, simply because it is much easier to obtain all the data for one city from one department. However, I think the chronological trends in melanoma are reliable, particularly in the mortality of young adults, where there is very unlikely to be either diagnostic or recording confusion. Furthermore, there have been a number of studies that have directly compared the pathological classification of tumours recognized 20 years ago, to determine whether there have been classification changes. Although these studies showed minor variations, they did not suggest that any major proportion of the large incidence trend over the last few decades is a reporting phenomenon. If there have been artefactual increases it more likely applies to the last decade when one's willingness to do a biopsy of a suspicious lesion has perhaps increased.

Q I was interested that the correlation between sun spot activity and ultraviolet did not actually work out in relationship to the incidence of melanoma, but the periodicity of sunspots suggest that there might be a correlation with age-specific incidence of melanoma, and the higher levels of ultraviolet that would operate in between the periods of maximum sunspot activity. Has that analysis been done?

Professor Elwood Perhaps not in the way that you are suggesting. One of the crucial things is to start with a very clear assumption as to what the latest period is likely to be because, if you just start analysing data, you can produce any number of possible correlations.

Dr Keith Shine Our data tell us that ozone level is correlated with solar cycles and that solar maxima produce about 2% more ozone than solar minima.

Professor Elwood Yes, that effect was put forward by Berger as the explanation for the decrease in ground-level ultraviolet associated with solar maxima.

Bob Watson In reality there is no variation in ultraviolet radiation from the sun above 240 nm. There is more ozone in the stratosphere because there is some variability around 200/240 nm and the phota dissociate the oxygen. Thus there is an inverse correlation with sunspot activity and ultraviolet reaching the ground, not a positive correlation.

Dr David Gould (Truro, Cornwall) Have you had the opportunity to look at any other immune-competent cells in patients with XP, and in particular have you had the chance to look at Langerhans cells?

Dr Colin Arlett Not personally. There have been reports of variability in the sensitivity or function of Langerhans cells in XP. We reviewed the literature on this recently and there is no consensus on this point. On the other hand, the reduction in NK activity seems fairly definitive.

Rona Mackie I think it does have to be said that there is huge diurnal variation in NK activity, and many other things do cause alterations.

Don Cussack (USDA Plant Stress Lab.) A number of years ago there was an effort in England to assess people's active ultraviolet dose. Did those studies generate any useful data?

Dr van der Leun Yes. There are films that you can wear as a badge, and these measure the ultraviolet dose you receive. The dose is then rated according to a sensitivity curve approaching that of human skin erythema. These badges were used in several investigations and yielded some interesting results. For instance, that outdoor workers receive roughly three times as much ultraviolet as indoor workers. Formerly we had to guess such numbers, but it is nice to know them now from actual measurements. One thing that surprised me is that the fraction of the ultraviolet available outdoors that actually reaches the skin hardly depends on the season. Office workers had I think 3 or 4% in summer and also 3 or 4% of the available dose in winter. This helps when making quantitative predictions.

Dr Wright (London Hospital) Some of the data presented by Dr Arlett suggested that reduced colony-forming ability of fibroblasts in cultures may act as a predictor of vulnerability of a patient to ultraviolet-induced skin cancer. Your data also showed that colony-forming ability was very tight among 'normals'. Is this true across races?

Dr Arlett We have not looked at any fibroblast cultures from deeply pigmented skin. As far as Caucasians are concerned, it is extremely predictable. There is no difference in the colony-forming ability of sun-sensitive and non-sun-sensitive Caucasians. I cannot give a formal answer as far as other races are concerned. The Japanese seem to follow exactly the same pattern, but of course the Japanese are not very pigmented. The sensitivity of fibroblasts from XP patients does not absolutely correlate with the extent of the defect in excision repair. So fibroblast culture does not help very much in predicting outcome for skin cancer. It is useful, however, in that the more sensitive cells correlate closely with neurological defects, particularly complementation groups A and D, so if a child is found to have very low ultraviolet survival it almost certainly means complementation group A or complementation group D and implies neurological defects by the age of 3.

Part 6
Global Consequences

15
Effects of Ultraviolet-B Radiation on Terrestrial Plants and Marine Organisms

ROBERT C. WORREST and LESTER D. GRANT
US Environmental Protection Agency

ABSTRACT

Experimental evidence suggests that increased exposure to UVB radiation (290–320 nm) at the Earth's surface would have negative effects on both terrestrial and aquatic biota. Despite uncertainties resulting from the complexities of field experiments, the data presently available indicate that crop yields are potentially vulnerable to increased levels of solar UVB radiation. Increased levels of UVB radiation may also affect forest productivity. Existing data also suggest that increased UVB radiation will modify the distribution and abundance of plants, and thereby change ecosystem structure. For components of marine ecosystems, various experiments have demonstrated that UVB radiation causes damage to fish larvae and juveniles, and other small animals and plants essential to the marine food web. Effects include decreases in reproductive capacity, growth, survival and other functions. Experimental evidence suggests that even small increases in ambient UVB exposure could result in significant ecosystem changes.

INTRODUCTION

Inadvertent alterations of the Earth's atmosphere by human activities are now of regional and even global proportion. In the last decade, scientists, policy-makers and the general public have focused increasing concern on the consequences of a

---

Ozone Depletion: Health and Environmental Consequences
Edited by R. Russell Jones and T. Wigley
© 1989 John Wiley & Sons Ltd

reduction of ozone in the upper atmosphere. The effects of this depletion will show no respect for national borders and will affect the entire Earth — a problem of truly global scale. If individual nations or the international community find stratospheric ozone reduction to be unacceptable, it could take decades to centuries to reverse the depletion. Thus, anticipation of the consequences of stratospheric ozone reduction and efforts to mitigate or adapt to such depletion are extremely important.

Based on a report from the International Ozone Trends Panel (which involved over one hundred scientists) formed by the National Aeronautics and Space Administration (NASA), National Oceanic and Atmospheric Administration (NOAA), Federal Aviation Administration (FAA), World Meteorological Organization (WMO), and the United Nations Environment Programme (UNEP), there is undisputed evidence that the atmospheric concentrations of source gases important in controlling stratospheric ozone levels continue to increase on a global scale because of human activities.[1] A reduction in ozone concentration will result in increased transmission of solar ultraviolet radiation through the stratosphere. Various experiments have identified many adverse, serious effects of such an increase in exposure to this radiation, and the effects will continue well into the next century even with a vigorous mitigation program. To establish responsible regulations and mitigation options, we need to know more precisely what the effects of ozone depletion are likely to be. The potential effects could seriously threaten the quality of life as we know it.

What follows is a summary of important issues and key scientific information concerning stratospheric ozone depletion, including background information on its causes and current or anticipated extent, and its potential adverse effects on the environment.

CAUSES AND EXTENT OF OZONE DEPLETION

Speculation regarding the possibility of stratospheric ozone reduction first appeared in the early 1970s and was focused on the consequences of large quantities of nitrogen oxides being injected into the upper atmosphere by supersonic aircraft flying at high altitudes (see Chapter 4, Figure 1). Researchers also considered other sources of nitrogen oxides originating from the Earth's surface. With further refinement, concerns about nitrogen oxide pollution of the upper atmosphere diminished because the quantities likely to be involved were insufficient to cause a serious threat to the ozone layer. However, in the mid-1970s, concern for halogen pollution of the upper atmosphere surfaced. The halogens of immediate concern were chlorine and bromine. The source for chlorine is primarily chlorofluorocarbons (CFCs), which are released worldwide from, among others, such sources as aerosol spray cans, certain plastic foams and refrigerative air conditioners. The sources for bromine are the halons used in fire extinguishers.

Many gases emitted as a result of the industrial and agricultural activities of humans can accumulate in the Earth's atmosphere and ultimately contribute to alterations in the vertical distribution and concentrations of stratospheric ozone. Among the most important are those trace gases having long residence times in the atmosphere, allowing for accumulation in the troposphere and their gradual upward migration into the stratosphere, where they contribute to depletion of stratospheric ozone. The atmospheric and chemical processes involved are extremely complex, as reviewed elsewhere (e.g. Watson et al.[1]). Trace gases of particular concern include:

(1) Certain long-lived chlorofluorocarbons, such as CFC 11, CFC 12, and CFC 113, that have atmospheric residence times of approximately 75 to 110 years.
(2) Methyl chloroform (CH_3CCl_3) and carbon tetrachloride (CCl_4), with approximate residence times of 8 and 67 years, respectively.
(3) Halon 1301 and halon 1211, with 110 and 25 year residence times, respectively.

Given the time involved in transport of these gases to the stratosphere, their long residence times in the atmosphere, and their slow removal processes, any effects already seen on stratospheric ozone are, in part, the result of atmospheric loadings due to anthropogenic emissions several decades ago. Those gases already in the atmosphere will continue to exert stratospheric ozone depletion effects far into the future – that is, well into the next century.

The atmospheric models that predict future ozone depletion are in a continual process of refinement. Through the years the predicted scenarios for resultant changes in stratospheric ozone have ranged from –4 to –18% for steady-state concentrations of chlorine consistent with recent levels of CFC 11 and CFC 12 emissions. However, over time we have come to realize that other gases will influence column ozone and that the magnitude and sign of the predicted change in total ozone over the next century depend critically on the multiple trace-gas scenarios that are assumed (see Chapter 4). A number of the modeling scenarios assume relatively uniform rates of ozone layer reduction, widely distributed above all regions of the Earth. However, scientists have identified areas of distinctly greater depletion (ranging to 50% in recent years) over the South Polar region during September to November of each year; and evidence suggests a likely gradual expansion of this Antarctic ozone hole ultimately to extend beyond the South Polar region, possibly reaching more heavily populated areas of the Southern Hemisphere. Similarly, it is considered likely that an analogous, though less intense, zone of upper level ozone reduction will ultimately occur over the North Polar region and will expand out over populated areas of the Northern Hemisphere as well. Preliminary results from the 1989 Airborne Arctic Stratospheric Expedition have substantially increased confidence in the scientific understanding of the polar stratospheric-cloud-induced, chlorine catalyzed, ozone-depletion phenomenon found in both the Arctic and the Antarctic.

Although ozone constitutes a very small proportion of the gaseous composition of the stratosphere, it plays a major role in protecting life on this planet. The result of changes in the density of the total ozone column could, therefore, be far-reaching. The natural distribution of ozone in the Earth's atmosphere, concentrated most heavily in a diffuse layer in the stratosphere, is crucial in helping to protect humans, other biological systems, and man-made materials from the harmful effects of certain wavelengths of sunlight. Stratospheric ozone exerts its beneficial effects by absorbing ultraviolet radiation in the 200–320 nm range, with reduced amounts of radiation in the 290–320 nm waveband (ultraviolet-B or UVB radiation) penetrating to the Earth's surface. Also, the vertical distribution of stratospheric ozone is a factor in maintaining the radiative balance of the Earth's atmosphere and, therefore, the temperature of the Earth. Depletion of the stratospheric ozone layer can, therefore, lead to damaging effects on the environment directly by increased penetration of UVB radiation to the Earth's surface and indirectly by the influences of changes in the vertical distribution of stratospheric ozone and water vapor that contribute to global warming effects and altered climatic conditions.

If a decrease in total atmospheric ozone were to occur, exposure to solar radiation in the 290–320 nm waveband would increase. The percentage increase in surface UVB radiation is a little more than double the percentage decrease in total column ozone. A UVB increase would have an effect on humans, other animals, plants, certain manufactured materials, and photochemical smog production. Most of the known biological effects of UVB radiation are damaging; therefore, the possibility of increased exposure to solar UVB radiation poses particular cause for concern. Detailed discussions of evolving concern about stratospheric ozone depletion and assessments of scientific bases underlying such concern can be found in several recent national and international expert working group reports or symposia.[2-4] The following sections summarize key ecological points from such sources and discuss their implications for development of effective international efforts either to mitigate or to cope with ozone layer depletion.

EFFECTS ON TERRESTRIAL PLANTS AND MARINE ORGANISMS

Terrestrial plants

Increased UVB irradiation of the Earth's surface due to ozone layer depletion can be expected to exert negative impacts on both terrestrial and aquatic biota. In asssessing the impact of increased exposure to UVB radiation for crops and terrestrial ecosystems it must be recognized that existing knowledge is, in many ways, deficient. The effects of enhanced levels of UVB radiation have been studied in only a few representative species from some of the major terrestrial ecosystems. Most of our knowledge is derived from studies focused on agricultural crops and conducted at mid-latitudes. Despite uncertainties resulting from the complexities of field experiments, the data currently available suggest that crop yields are potentially vulnerable to increased levels of solar UVB radiation. Unlike drought or other geographically isolated stresses, stratospheric ozone depletion would affect all areas of the world, including ecosystems whose UVB sensitivity has not been investigated.

Out of more than 200 species and cultivars screened for ultraviolet tolerance, experiments have shown about two thirds to be sensitive. Over 90% of the species examined have been crop plants; virtually nothing is known about the responses of nonagricultural, terrestrial plants to UVB radiation. Most tests were carried out in controlled environments with ultraviolet radiation from artificial sources and where ultraviolet sensitivity is amplified when compared with results obtained by exposure to solar radiation in the field. The most sensitive plant groups include crops related to peas and beans, melons, mustard and cabbage. Of the crops studied in the field, there are still large differences in sensitivity.[5] In general, ultraviolet radiation causes reduced leaf and stem growth, lower total dry weight, and lower photosynthetic activity in sensitive cultivars of plant species.[6] These results were corroborated in an experiment simulating a 25% enhancement of solar UVB radiation (equivalent to 12% ozone reduction), where UVB exposure was controlled by an artificial ozone filter at a high elevation at a southern latitude.[7] Members of the grass family were generally less sensitive (with some notable exceptions), possibly due to protective abilities such as photorepair or production of screening pigments.[8]

The high degree of variation in sensitivity that exists among cultivars within each crop species suggests that some degree of ultraviolet tolerance must be present in the existing gene pool. The genetic basis for differences in UVB sensitivity is not fully understood; however, there is a possibility that selective crop breeding may help mitigate some of the potentially deleterious effects.[9]

In addition to other factors, increased levels of UVB radiation may reduce the quality of crop yield. Few studies have specifically examined changes in crop quality, but studies of certain cultivars of tomato, potato, sugar beets and soybean have noted reduced quality. The protein and oil content of specific cultivars of soybean seeds were reduced by up to 10% when plants were exposed to ultraviolet levels approximating a 25% ozone depletion.[5]

Increased levels of UVB radiation may also affect forest productivity. Only limited data are available on coniferous species, but UVB radiation adversely affected about one half of the species of seedlings studied.[10] In loblolly pine seedlings, growth and photosynthesis were reduced in field studies simulating a 40% ozone reduction.[11] However, extrapolation from results of seedling studies to forested ecosystems is not possible, nor is interpolation of predicted results at exposure levels simulating less ozone reduction.

Existing data also suggest that increased UVB radiation will modify the distribution and abundance of plants, and potentially change ecosystem structure as a result of an alteration of the competitive balance between different species. Even small changes in competitive balance over a period of time can result in large changes in community structure and composition.[12] The shift in competitive balance may occur in response to subtle changes in plant growth, without large changes in fundamental physiological processes such as photosynthesis.[13] The alteration of the competitive balance of species is a dynamic process determined by the competing species and their immediate environment. Unfortunately, the current knowledge base provides neither a quantitative nor a qualitative prediction of how these ecosystems might be altered.

Aquatic organisms

For components of marine ecosystems, various experiments have demonstrated that UVB radiation causes damage to fish larvae and juveniles, shrimp larvae, crab larvae, and other small animals and plants essential to the marine food web. Effects include decreases in reproductive capacity, growth, survival and other functions.[14,15] Although not nearly as important as light, temperature or nutrient levels, evidence indicates that ambient solar UVB radiation is currently an important limiting ecological factor, and that even small increases of UVB exposure could result in significant ecosystem changes.[16]

Effects induced by solar UVB radiation have been measured to a depth of more than 20 m in clear waters and more than 5 m in unclear water. The euphotic zone (i.e. those depths with levels of light sufficient for photosynthesis) is frequently taken as the water column that reaches down to the depth at which photosynthetically active radiation is reduced by 99%. In marine ecosystems, UVB radiation penetrates approximately the upper 10% of the marine euphotic zone before it is reduced 99% from its surface irradiance. Penetration of UVB radiation into natural waters is a key variable in assessing the potential impact of this radiation on any aquatic ecosystem.[15]

In marine plant communities a major change in species composition — in addition to a global decrease in net production — might result from enhanced UVB exposure.[17] A change in community composition at the base of food webs may produce instabilities within ecosystems that would affect higher trophic levels.[18] The generation time of marine phytoplankton is in the range of hours to days; whereas the potential increase in ambient levels of solar UVB irradiance will occur over decades. The question remains as to whether the gene pool within species is flexible enough to adapt during this relatively gradual (relative to the generation time of the target organisms) change in exposure to UVB radiation. There is evidence that a decrease in column ozone abundance could diminish the near-surface season of invertebrate zooplankton populations. For some zooplankton, the time spent at or near the surface is critical for food gathering and breeding. Whether the population could endure a significant shortening of the surface season is unknown.[19]

The direct effect of UVB radiation on food-fish larvae closely parallels the effect on invertebrate zooplankton. Before effects of exposure to UVB radiation can be predicted, researchers require information on seasonal abundances and vertical distributions of fish larvae, vertical mixing, and penetration of UVB radiation into appropriate water columns. However, in one study involving anchovy larvae, it was calculated that a 20% increase in UVB radiation (which would accompany a 9% depletion of total column ozone) would result in the death of about 8% of the annual larval population.[20] This one study was performed in the laboratory, and even the control animals had significant mortality at the end of the normal larval period. This highlights the need for caution when trying to extrapolate conclusions based on results from laboratory studies to natural conditions.

More than 30% of the world's animal protein for human consumption comes from the sea. In many developing countries, this percentage is even larger. We need research to improve our understanding of how stratospheric ozone depletion could influence the world food supply. Past efforts to understand the effects of stratospheric ozone depletion on marine ecosystems have addressed the physical variables that determine the exposure (or dose) to the systems, as well as the amount of UVB radiation that produces biological effects, the direct effects on the sensitive life stages of ecologically important marine organisms, and the extent to which the sum of the effects might affect the marine resources significant to man. The sum of past efforts suggests that the likelihood of a significant impact resulting from UVB exposure of marine environments is real, even if it cannot yet be fully quantified.

CONCLUSIONS

The information summarized in the preceding sections indicates that stratospheric ozone depletion has the potential to exert very substantial effects on the environment. However, our ability to predict with confidence and precision the likelihood of particular effects occurring and to quantify their anticipated ultimate scope varies greatly due to differences in the current level of our state of knowledge about them. A major dilemma is the fact that our current state of knowledge is low with regard to certain effects that have the greatest potential for widespread global impacts. For example, the current state of knowledge concerning potential effects of increased UVB radiation on crop productivity is relatively low, but the impact on the global food supply could be quite high. The international scientific community, under the review provisions of the Montreal Protocol on Substances that Deplete the Ozone Layer, is currently preparing an assessment of the environmental effects of stratospheric ozone depletion. This assessment is scheduled for completion in August 1989, and will serve as input to the 1990 policy review of the Montreal Protocol.

It is clear that enhancements of chemically active chlorine compounds do occur in both the Arctic and Antarctic stratosphere, and represent an additional ozone-depleting process that was not included in the stratospheric ozone assessment models used as a basis for the Montreal Protocol. The definition of appropriate steps to be taken to deal with stratospheric ozone depletion has been the subject of discussion at numerous recent international symposia and work groups.[3,4] The general international consensus emerging from these meetings is that effective international cooperation is necessary to reduce future stratospheric ozone depletion and that implementation of the Montreal Protocol is urgently needed as a first important international cooperative step. However, it is clear that even if the Montreal Protocol is fully implemented as currently written, with worldwide participation, it will still allow atmospheric chlorine and bromine to increase by about a factor of two by the middle of the next century. Emission cuts of at least 80–90% worldwide are necessary to stabilize halogen concentrations. However, the currently available maximum control potential is expected to be below 70%. Even with a complete phase-out of the fully halogenated CFCs, full recovery of the ozone layer will take many decades to centuries given the long atmospheric residence times of CFCs. Ecological and biological research is required to determine what mitigative or adaptive strategies might be required to cope with the anticipated, unavoidable loss of stratospheric ozone.

The global community will need to follow several control approaches to accomplish the goals set by the Montreal Protocol. The first priority is to find, as substitutes for the CFCs already implicated in ozone destruction, chemicals with lower ozone depleting potential. A number of substitutes have already been suggested. In addition, products and equipment have to be developed that require less or no CFCs. Also, the technology for emission control in production processes, as well as recovery and recycling of used CFCs, need further development.

In summary, to reduce the impact of future stratospheric ozone depletion, the industrialized and developing countries need to take immediate steps, both domestically and internationally, to reduce use of CFCs and halons with long atmospheric residence times and strong stratospheric ozone depletion potential. Also, the scientific community needs to produce baseline ecological and biological information regarding the effects of solar UVB radiation – information that can be used to develop mitigative and adaptive strategies to address the unavoidable loss of stratospheric ozone.

Although compilation of the information contained within this chapter has been funded in part by the US Environmental Protection Agency, the chapter has not been subjected to Agency review and therefore does not necessarily reflect the views of the Agency and no official endorsement should be inferred.

EFFECTS OF UVB ON PLANTS AND MARINE ORGANISMS

REFERENCES

1. R. T. Watson, Ozone Trend Panel, M. J. Prather, *Ad Hoc* Theory Panel, M. J. Kurylo, and *NASA* Panel for Data Evaluation, *Present State of Knowledge of the Upper Atmosphere 1988: An Assessment Report*, National Aeronautics and Space Administration, NASA Reference Publication 1208, Washington DC (1988).
2. US Environmental Protection Agency, *Assessing the Risks of Trace Gases that Can Modify the Stratosphere*, Vols. I–V. EPA/Office of Air and Radiation, USEPA/400/1-87/001A-E, Washington DC (1987).
3. T. Schneider, S. D. Lee, G. Wolters and L. D. Grant, eds., *Atmospheric Ozone Research and Its Policy Implications*, Proceedings of the 3rd US–Dutch International Symposium, Nijmegen, The Netherlands. Elsevier, Amsterdam (1988).
4. World Meteorological Organization and Canada Department of the Environment, *Proceedings of the World Conference on the Changing Atmosphere*, Toronto (1988).
5. US Environmental Protection Agency, Risks to crops and terrestrial ecosystems from enhanced UV-B radiation. In J. Hoffman, ed., *Assessing the Risks of Trace Gases that Can Modify the Stratosphere*, pp (11)1–31, USEPA 400/1-87/001C, Washington DC (1987).
6. M. Tevini and W. Iwanzik, Effects of UV-B radiation on growth and development of cucumber seedlings. In R. C. Worrest and M. M. Caldwell, eds, *Stratospheric Ozone Reduction, Solar UV Radiation and Plant Life*, Springer-Verlag, Heidelberg (1986).
7. M. Tevini, D. Steinmüller and W. Iwanzik, *Uber die Wirkung erhöhter UV-B-Strahlung in Kombination mit anderen Stressfaktoren auf Wachstum und Funktion von Nutzpflanzen*. BPT-Bericht 6/86, GSF Munich (1986).
8. C. J. Beggs, U. Schneider-Ziebert and E. Wellmann, UV-B radiation and adaptive mechanisms in plants. In R. C. Worrest and M. M. Caldwell, eds, *Stratospheric Ozone Reduction, Solar UV Radiation and Plant Life*, Springer-Verlag, Heidelberg (1986).
9. A. H. Teramura, Effects of ultraviolet-B radiation on the growth and yield of crop plants. *Physiol. Plantarum*, **58**, 415–427 (1983).
10. J. Sullivan and A. H. Teramura, The effects of ultraviolet-B radiation on seedling growth in the Pinaceae. *Am. J. Botany*, **75**, 225–230 (1988).
11. A. H. Teramura and J. Sullivan, *Annual Report to the US Environmental Protection Agency: The Effects of Changing Climate and Stratospheric Ozone Modification on Plants*. University of Maryland, College Park, Maryland (1988).
12. W. G. Gold and M. M. Caldwell, The effects of ultraviolet-B radiation on plant competition in terrestrial ecosystems. *Physiol. Plantarum*, **58**, 435–444 (1983).
13. W. Beyschlag, P. W. Barnes, S. D. Flint and M. M. Caldwell, Enhanced UV-B irradiation has no effect on photosynthesis characteristics of wheat (*Triticum aestivum* L.) and wild oat (*Avena fatua* L.) under greenhouse and field conditions. *Photosynthetica*, **22**, 31–37 (1988).
14. R. C. Worrest, Review of literature concerning the impact of UV-B radiation upon marine organisms. In J. Calkins, ed., *The Role of Solar Ultraviolet Radiation in Marine Ecosystems*, pp 429–458, Plenum Press, New York (1982).
15. US Environmental Protection Agency, An assessment of the effects of ultraviolet-B radiation on aquatic organisms. In J. Hoffman, ed., *Assessing the Risks of Trace Gases that Can Modify the Stratosphere*, pp (12)1–33, USEPA 400/1-87/001C, Washington DC (1987).
16. D. M. Damkaer, Possible influence of solar UV radiation in the evolution of marine zooplankton. In J. Calkins, ed., *The Role of Solar Ultraviolet Radiation in Marine Ecosystems*, pp 701–706, Plenum Press, New York (1982).

17. R. C. Worrest, Impact of solar ultraviolet-B (290–320 nm) radiation upon marine microalgae. *Physiol. Plantarum*, **58**, 428–434 (1983).
18. J. R. Kelly, How might enhanced levels of solar UV-B radiation affect marine ecosystems? In J. G. Titus, ed., *Effects of Changes in Stratospheric Ozone and Global Climate*, US Environmental Protection Agency and United Nations Environment Programme, Washington DC (1986).
19. D. M. Damkaer, D. M. Dey, G. A. Heron and E. F. Prentice, Effects of UV-B radiation on near-surface zooplankton of Puget Sound. *Oecologia*, **44**, 149–158 (1980).
20. J. R. Hunter, S. E. Kaupp and J. H. Taylor, Assessment of the effects of UV radiation on marine fish larvae. In J. Calkins, ed., *The Role of Solar Ultraviolet Radiation in Marine Ecosystems*, pp 459–497, Plenum Press, New York (1982).

16
Consequences for Human Health of Stratospheric Ozone Depletion

ROBIN RUSSELL JONES
St John's Hospital for Diseases of the Skin, London, UK

The absorption of ultraviolet light by stratospheric ozone is crucial to the protection of living organisms. UVC in the 240–290 nm range is virtually eliminated by ozone, and only a proportion of UVB (290–320 nm) penetrates the terrestrial environment. Because UVB and UVC span the photo-absorption spectrum of DNA, ozone is crucial to the viability of micro-organisms and other primitive life forms. UVC is particularly damaging to DNA; a property exploited in germicidal lamps. Figure 1 shows DNA damage increasing by four orders of magnitude between 320 and 280 nm, reaching a maximum at 265 nm, well into the UVC range. The mantle of stratospheric ozone that encompasses planet Earth ensures that very little UVC reaches the surface of the planet and therefore provides an environment suitable for the evolution of terrestrial life. Predicting the consequences for human health of reversing that process has received considerable attention,[1-4] and is the subject of this review.

The important effects on human health associated with stratospheric ozone depletion are summarized in Table 1. I will concentrate mainly on the direct health effects of ultraviolet radiation, since this is where most of the scientific certainty lies and the likely increases are relatively easy to predict. However, UVB irradiation depresses the immune response in mammalian species. Activation of the herpes virus leading to cold sores commonly occurs at the beginning of a summer holiday, and this reflects the immunosuppressive effect of ultraviolet light on human skin. Susceptibility to more important cutaneous infections such as leishmaniasis or leprosy may also be increased by exposure to ultraviolet light,

---

Ozone Depletion: Health and Environmental Consequences
Edited by R. Russell Jones and T. Wigley
© 1989 John Wiley & Sons Ltd

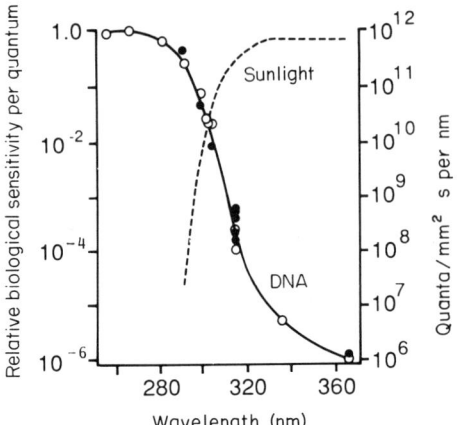

Figure 1 The continuous line represents an average action spectrum for DNA damage (adapted from Setlow[50]. The dotted line represents solar flux at the Earth's surface.

Table 1 Potential health effects of stratospheric ozone depletion

Direct effects of UVB
Acute
  Acute erythema (sunburn)
  Acute keratitis (snow-blindness)

Chronic
  Cataracts
  Skin cancer

Indirect effects of UVB due to decreased immunosurveillance
  Increased susceptibility to cutaneous infection
  Carcinogenesis

since the behaviour of these conditions is closely linked to cell-mediated immunity in the skin.[5] Finally, it is known from experimental work that transplantation of tumours into animal models is enhanced if the subject has been previously exposed to ultraviolet radiation.[6,7] Maximal affect is obtained with light in the 290–300 nm range,[8] so again stratospheric ozone depletion may have more general effects on carcinogenesis than those associated with direct DNA damage.

Information on the link between sunlight and skin cancer comes from three separate sources: clinical observation, epidemiological studies and experimental data. Perhaps the most important clinical observation is that skin cancer risk is determined by skin colour, and that black skin is virtually immune to the development of ultraviolet-related skin cancers. It has been traditional to divide skin types into six categories according to their response to sun exposure and the amount of pigmentation present in their skin (Table 2).[9] Melanin is an effective absorber of ultraviolet light, and Figure 2 shows the relative transmission of ultraviolet light in white and black skin.[11] The difference is remarkable, particularly in the UVC and UVB range.

Rather like ozone, melanin virtually blocks the transmission of ultraviolet light below 300 nm, a range which is known from experimental data to be of critical importance in the induction of skin cancer.

The second important observation is that in white populations there is an inverse relationship with latitude and the incidence of three types of skin cancer: basal cell carcinoma, squamous cell carcinoma and malignant melanoma.[12,13] Figure 3 plots annual age-adjusted incidence rates (logged) in white males for these three cancers against biologically effective ultraviolet dose at different latitudes in North America. All three cancers show a rising incidence with

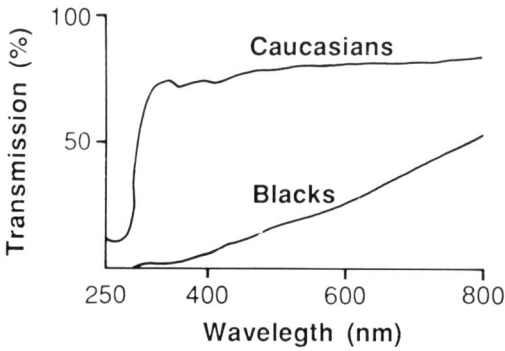

Figure 2 Comparative transmittance of ultraviolet radiation in white and black skin. (Reproduced from S. Wan et al.[10])

Table 2 Classification of skin type

1	Always burn	Never tan
2	Always burn	Sometimes tan
3	Sometimes burn	Always tan
4	Never burn	Always tan
5		Brown
6		Black

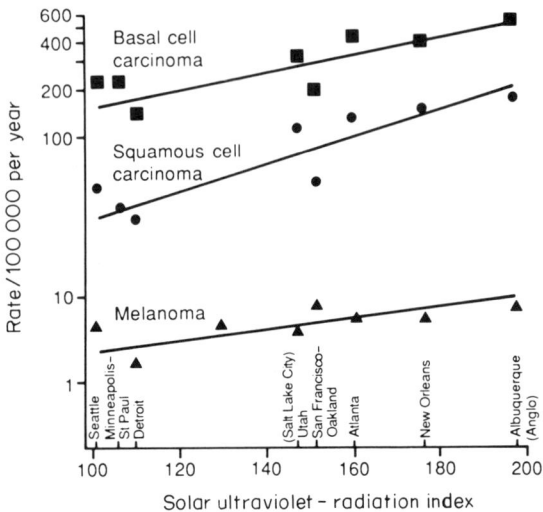

Figure 3 Annual age-adjusted incidence rates for BCC, SCC and CMM among white US males. (Source: Environmental Protection Agency[1] Chapter 7, 1977–78.)

increasing dose, but the relationship is stronger for squamous cell carcinoma than for basal cell carcinoma or malignant melanoma. This figure also demonstrates that basal cell carcinoma is commoner than squamous cell carcinoma, the normal ratio being about 4 : 1. Basal cell and squamous cell carcinoma are commonly grouped together as non-melanoma skin cancer (NMSC). Figure 4 plots the incidence of NMSC in eight countries of different proximity to the equator.[14] Again a linear relationship is obtained when log incidence of NMSC is plotted against annual ultraviolet solar radiation. There are, however, important clinical differences between these types of non-melanoma skin cancer, and they will therefore be reviewed separately.

BASAL CELL CARCINOMA

Apart from its latitude dependence, basal cell carcinoma (BCC), also called basal cell epithelioma or rodent ulcer, is linked to ultraviolet exposure by a number of clinical and epidemiological observations:[14-16]

(1) It is commoner in fair-skinned individuals and virtually unknown in people of Negro extraction, unless they happen to be albino.
(2) The incidence of BCC increases with age.
(3) The tumours usually arise in areas of chronically sun-damaged skin, and are associated histologically with elastosis.

(4) 90% of tumours arise on the head or neck.
(5) The incidence of BCC is higher in outdoor workers, such as farmers or fishermen.
(6) The incidence of BCC is high in patients with xeroderma pigmentosum, an autosomal recessive genodermatosis characterized by premature ageing of the skin and specific repair defects of ultraviolet-induced DNA damage.[17] By the age of 15, almost half the patients with xeroderma pigmentosum have developed either a basal cell carcinoma or a squamous cell carcinoma.[18]

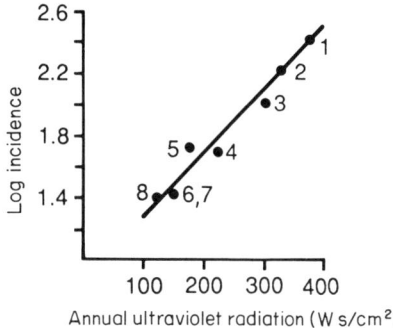

Figure 4 Logarithms of incidence rates for non-melanoma skin cancer per 100 000 males against annual ultraviolet solar radiation in watt seconds per cm^2. Male populations: 1, Queensland; 2, Texas; 3, South Africa (Cape Province whites); 4, Nevada; 5, Canada (six provinces); 6, England (five regions); 7, German Democratic Republic; 8, Scotland. (Reproduced from Gordon and Silverstone[14].)

However, the anatomical distribution of basal cell carcinoma indicates that ultraviolet light is not the only factor that determines tumour development. Thus, BCC seldom develops on the dorsal surface of the hands, a site of high ultraviolet exposure, whereas they are commonly found in a periorbital or retroauricular location. Furthermore, there is no experimental model for the induction of BCC in animals by ultraviolet light, and this limits our ability to define the ultraviolet action spectrum for BCC induction. Even so, the weight of clinical and epidemiological evidence supports the role of cumulative ultraviolet exposure in the aetiology of this cutaneous malignancy.

SQUAMOUS CELL CARCINOMA

Squamous cell carcinoma (SCC), also called squamous cell epithelioma, has many features in common with basal cell carcinoma. Thus, the incidence increases with age, it is commoner in fair-skinned individuals, outdoor workers, and patients with xeroderma pigmentosum. In white populations, 70% of lesions occur on the head and neck, whilst in black populations this site is rarely involved. Solar keratoses are precursor lesions, and the tumours arise in areas of chronically sun-damaged skin and are associated histologically with elastosis.[14–16]

In other ways the relationship of SCC with ultraviolet light is stronger than for BCC. Thus there is a steeper latitudinal gradient, sites with high ultraviolet exposure are not spared, and sites without ultraviolet exposure are seldom affected in the absence of some other carcinogenic insult. Finally, ultraviolet light can initiate and promote the production of SCC in animal models. Indeed, our knowledge of the carcinogenic impact of different wavelengths of ultraviolet light is based largely on such experimentation.

In conclusion, there is incontrovertible evidence of the relationship between chronic ultraviolet exposure and the incidence of squamous cell carcinoma. At the same time, it should be recognized that more than one mechanism may be involved in this process. Immune surveillance is critical for the recognition and elimination of cells with malignant potential. In the case of squamous cell carcinoma, it is believed that antigen-presenting cells within the epidermis, Langerhans cells, recognize keratinocytes that have undergone mutations, probably because their cytokeratin profile is antigenically different from the cytokeratins normally found in the host epidermis.[22] SCC production is accelerated in situations in which the immune system is compromised, either because of myeloproliferative diseases such as leukaemia, or through the administration of immunosuppressive drugs.[23–25] This effect is particularly well seen in the recipients of organ transplants, where immunosuppressive therapy may need to be continued for decades. The change in incidence is greatest for SCC and least for BCC, with cutaneous malignant melanoma (CMM) occupying an intermediate position. In one series of transplant patients, BCC incidence did not alter significantly, CMM was elevated five times, and SCC 40 times.[26] The normal BCC : SCC ratio of 4 : 1 is therefore reversed in organ transplant recipients. However, this effect is latitude-dependent, with larger SCC : BCC ratios in areas of high ultraviolet exposure.[27] Table 3 shows an SCC : BCC ratio ranging from 1.3 : 1 in Minnesota to 16 : 1 in New South Wales, Australia.

These considerations are also relevant to the mechanism of ultraviolet carcinogenesis, since ultraviolet light is also a potent immunosuppressant. For the induction of SCC, both direct DNA damage and immune surveillance are

Table 3 SCC : BCC ratios in organ transplant recipients at different locations

Minnesota	1.3 : 1
Toronto	2.3 : 1
Sydney	11 : 1
Denver, Colorado	11 : 1
New South Wales	16 : 1

clearly important. Less certain is how these two mechanisms interact, and whether they are of equal importance in other types of ultraviolet-induced skin cancer.

MALIGNANT MELANOMA

It is not necessary to consider in detail the epidemiology of cutaneous malignant melanoma (CMM), since this is covered elsewhere (see Chapter 14).

Again, however, there is a clear latitude dependence for the three commonest varieties of melanoma: superficial spreading melanoma (SSM), nodular melanoma (NNM) and lentigo maligna melanoma (LMM), but not for acral lentiginous melanoma (ALM). ALM arises in non-ultraviolet-exposed sites, such as the palms, soles or sub-ungual areas, and is the only melanoma found in people who have never lived south of the Arctic Circle.[28]

Age-specific incidence rates for CMM are quite different from those for BCC and SCC, except for LMM, which increases in frequency with age, arises in chronically sun-damaged skin, and is associated histologically with elastosis.[29]

Melanoma frequency is considerably increased in patients with xeroderma pigmentosum, being approximately 2000 times greater than normal in a 20-year-old population.[30] The incidence of malignant melanoma is rising in all Caucasian populations studied, and the mean age of diagnosis is falling.[31-33] CMM is particularly common in social class 1[34] and fair-skinned individuals who burn easily.[33] SSM and NM are the only skin tumours in which working indoors increases risk; an effect which is independent of social class.[35,36]

Whilst debate continues about the relative importance of different risk factors in CMM, it is generally agreed that short periods of high ultraviolet intensity are of greater significance than cumulative ultraviolet exposure.[37] Furthermore, melanoma epidemiologists can make some fairly uncompromising predictions as to the results of different sunbathing habits. For example, girls who wear bikinis or sunbathe nude in early adult life carry a risk of developing melanoma on the trunk 13 times greater than those who wear one-piece swimsuits.[38] Skin type is also crucial. Adolescents with fair skin, fair hair and numerous freckles carry a risk of melanoma 37 times greater than those with dark skin, dark hair and few freckles.[35]

However, our knowledge of melanoma induction is limited by the lack of suitable experimental data. Animal models are being developed, but it has not yet been possible to define an action spectrum for melanoma induction. Furthermore, predisposition to melanoma through familial conditions such as dysplastic naevus syndrome is clearly an important factor, which needs to be incorporated in future melanoma surveys.[39] Because of this uncertainty and the complexity of the relationship between ultraviolet light and melanoma, our ability to predict the consequences of ozone depletion is reduced.

EPIDEMIOLOGY

Incidence rates for CMM are between 2 and 10% of those for NMSC, yet melanoma incidence assumes considerable public health significance due to the younger age group affected and the higher mortality rate. Thus, although CMM has an overall mortality rate of 25%, the highest figure claimed for NMSC is 1–2%,[40] and even this may be too high. In general melanoma statistics are considerably more reliable than those for NMSC. In most countries, including the UK, NMSC is not even included in the cancer registry statistics. Even so, estimates can be made based on surveys carried out in selected areas. The most comprehensive have been undertaken in the USA, where it is calculated that approximately 400000 new cases of NMSC develop annually. A mortality rate of 1–2% has been used for NMSC,[1,40] yielding up to 6000 deaths per annum in the USA. However, four out of five deaths attributable to NMSC result from SCC,[41,42] so this would yield an SCC mortality rate of around 6%. This figure seems high compared with Australian surveys, which indicate that only 1 in 60 SCC metastasize and that the mortality rate is probably less than 1%.[43,44]

By contrast the 25 800 cases of malignant melanoma seen annually in the USA result in 5800 deaths each year, a figure at least equivalent, and probably greater, than that for NMSC.

Table 4 shows incidence rates for melanoma and NMSC in different racial groups. In the USA the incidence of NMSC is 232 cases per 100 000 population per year in whites and 3.4 in blacks, a ratio of 60 : 1.[12,45] In the Bantu of South Africa the incidence rate of NMSC is said to be even lower (0.5 cases per 100 000 population per year), though the quality of the cancer registry data in this population may not be ideal.

Table 4 Estimated changes in UVB radiation in San Francisco using different weighting functions

	Ozone depletion	
	2%	10%
RB meter	1.6	8.6
Human erythema	3.5	19
Setlow DNA	4.3	23

Source: Environmental Protection Agency,[1] Vol. 3, Chapter 7.

For malignant melanoma the incidence rates for white US males in 1983 was 9.6 per 100 000 population per year, which is 19 times the rate in black US males. The female ratio was somewhat lower (8 : 1).[46]

It is also important to realize, however, that the skin cancers seen in black races may result from carcinogenic insults other than ultraviolet light. In Central Africa, for example, most SCC arises on the lower limb, and is probably the result of chronic trauma and ulceration. Similarly, melanoma expression in black populations is virtually confined to acral lentiginous melanoma, a form of melanoma that is not ultraviolet related.

It is clear, therefore, that the vast majority of ultraviolet induced skin cancers are seen in white populations, and that risk is increased by skin type and proximity to the equator. Darwinian principles have, of course, coped extremely well with this problem. Before the days of large-scale immigration, the skin colour of indigenous races was suited to the latitude of the world in which they evolved. Colonization of other lands, mainly by Europeans, has by-passed the normal evolutionary process and brought fair-skinned races in closer proximity to the equator than their degree of melanization ever intended them to be. When this is combined with an outdoor lifestyle, the results can be truly remarkable. The most recent survey of NMSC in Australia, which was based on a questionnaire sample of 30 000 of the population, revealed an overall incidence rate among white Australians of 823 per 100 000 population per year.[47] The incidence rates for all other cancers in Australia is 246, so NMSC is more than three times commoner than all other cancers combined. It also means that by the age of 75, two out of three Australians will have been treated for skin cancer.

Those resident in the North of Australia are at greater risk than those living in the South (incidence 1242 at latitudes <29°S compared with 489 per 100 000 per year at latitudes >37°S).

Australian-born whites are twice as likely to develop NMSC as British migrants (incidence 936 *versus* 402 per 100 000 per year), and those with skin type 1 showed a rate almost three times higher than those with skin type 4 (1764 *versus* 616 per 100 000 per year).

Higher rates were found in men than in women. Skin type 1 men showed the highest rate of all; 2140 non-melanoma skin cancers per 100 000 population per year.

The highest rate of malignant melanoma in the world is also found in Australia. In Queensland the incidence of malignant melanoma is 39.6 per 100 000 per year, which is ten times higher than in the UK, and four times higher than in the USA.[48] The lifetime risk of developing malignant melanoma for a white man born in Queensland is greater than 1 in 40, and the chances of dying of malignant melanoma approximately 1 in 150.

In conclusion, ultraviolet light can produce considerable mortality and morbidity through the induction of melanoma and non-melanoma skin cancers in white populations who live too close to the equator. The effect of ozone depletion will be to shift whole populations closer to equatorial conditions of ultraviolet exposure than their genetic constitution allows.

QUANTIFICATION OF BIOLOGICAL EFFECTS

Non-melanoma skin cancer

Quantifying the increase in skin cancer that will result from ozone depletion is a two-stage process that must take account in the increase of biologically effective ultraviolet light that results from an ozone loss of 1% (optical amplification factor, OAF), and the percentage increase in skin cancer that results from each increment in annual ultraviolet dose (biological amplification factor, BAF).[49] The value of BAF will vary according to the skin cancer under consideration and is derived from epidemiological data that correlate skin cancer incidence with geographical latitude. The OAF value depends on the absorption spectrum of ozone. It also depends on the system used to determine the biological effect of ultraviolet light at different wavelengths. Action spectra based on ultraviolet-induced DNA damage yield higher OAF values than those based on the induction of erythema in animal models or human skin. Setlow, for example, derived an OAF value of 2 based on the photo-absorption spectrum of DNA.[50]

A third method of weighting ultraviolet dose is the Robertson–Berger (RB) meter. This employs a magnesium tungstate sensor to monitor radiation in the UVB range, which is then weighted according to an action spectrum based on skin erythema in humans, but attaches greater weight to longer wavelengths.[51] A network of RB meters has been established at various national weather centres in the USA since 1974, and provides counts of weighted ultraviolet dose in sunburn units.[52]

Cole et al. have investigated the ability of different systems to predict the tumorigenic capacity of ultraviolet light using hairless mice as a model (Figure 5).[53] These authors concluded that the DNA action spectrum tended to overestimate, while RB meters tended to underestimate the carcinogenic impact of shorter wavelengths. More recently, Sterenborg and Slaper have more completely defined the action spectrum for the induction of SCC in hairless mice (see Chapter 13). Maximal effect in terms of carcinogenesis is obtained at 300 nm, falling off rapidly with increasing wavelengths. An OAF value of 1.6 has been derived from these data.

Use of these different action spectra produces different values for OAF. The NASA UVB model has been used to compare percentage changes in UVB radiation at various locations using different weighting functions, and the estimated changes are greater using human erythema or DNA action spectra than using RB meters (see Table 4). In practical terms it is generally agreed that a 1% loss of total column ozone results in an increase of 1.5–2.0% in the weighted ultraviolet flux, and that this approximation is acceptable for most heavily populated areas of the globe.

Calculating the value of BAF is complicated by the different skin tumours (SCC or BCC), the different values in male and female populations, and the fact that two different mathematical functions have been used to fit the

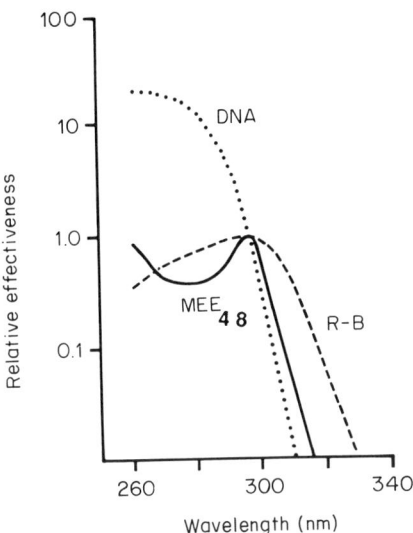

Figure 5 Action spectra comparing the RB meter with DNA damage and mouse oedema (MEE[48]). (Reproduced from Cole et al.[53])

epidemiological data.[54] The exponential model is a log–linear relationship and can be written

$$\ln R = \ln a + bU + e$$

where R is the skin cancer incidence, U the ultraviolet dose, a a constant, e an error term, and b the BAF.

The power law model is a log–log relationship and can be written

$$\ln R = \ln a + b \ln U + e$$

Using an exponential function, the value of BAF varies with latitude, being greater at lower latitude. A power-law relationship provides a constant value for BAF. In both cases the BAF value is independent of age, higher in men than in women, and higher for squamous cell carcinoma than for basal cell carcinoma. Table 5 details BAF values for BCC and SCC based on an analysis of the 1977–78 NCI survey using an exponential function,[55] and Table 6 details BAF values for NMSC based on two separate surveys of skin cancer in the USA using both exponential and power law functions.[54] The 1971–72 Third National Cancer Survey yields higher BAF values than the 1977–78 NCI Survey. Using a power-law relationship the figures are 2.7 *versus* 1.9 for men and 2.2 *versus* 1.4 for

Table 5 BAF values for BCC and SCC. Based on 1977–78 NCI Survey

	BCC	SCC
Men	1.32–2.59	2.08–4.09
Women	1.06–2.07	2.18–4.30

Table 6 BAF values for NMSC

	Power law	Exponential
1977–78 NCI Survey		
Men	1.9	1.4–2.3
Women	1.4	1.1–1.8
1971–72 TNC Survey		
Men	2.7	2.1–3.3
Women	2.2	1.7–2.7

women.[54] BAF values for basal and squamous cell carcinoma are based on the later NCI Survey and therefore provide lower estimates. The figure for SCC is approximately 3 and for BCC the figure is 2 for men and 1.5 for women. On this basis and using an OAF value of 2, a 10% reduction in total column ozone will increase the incidence of basal cell and squamous cell carcinoma by approximately 35 and 60%, respectively. In the USA this will produce an extra 160 000 cases of NMSC annually, and in the UK perhaps 8000.

The US Environmental Protection Agency has computed the effect of NMSC in the US population using six different emission scenarios for chlorine- and bromine-bearing substances.[1] These scenarios produce changes ranging from an increased ozone abundance of 2% by the year 2050 to an average global loss of 50% (Figure 6). A 10% loss will produce an extra 2 million cases of NMSC in the existing US population and for a cohort born between 2030 and 2074 an additional 92 million cancers will result (Table 7).

Cutaneous malignant melanoma

Attempts have also been made to derive formulae that will fit the epidemiological data for melanoma. Necessarily these are more complex due to the age distribution of malignant melanoma in Caucasian populations. An exhaustive analysis of the available data by the EPA concluded that for each 1% ozone depletion the incidence of CMM would increase by 1–2%.[1] However, there are no good experimental data that can be used to define the action spectrum for melanoma induction. After the Second World War changes in melanoma incidence coincided with changes in recreational activities, so it is likely that the

OZONE DEPLETION AND HUMAN HEALTH

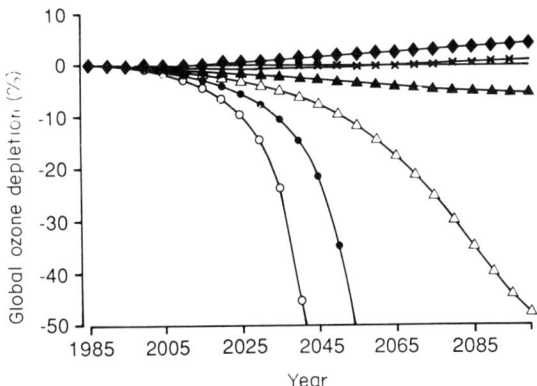

Figure 6 Global average ozone depletion based on six emission scenarios for chlorine- and bromine-bearing substances. (Source: Environmental Protection Agency,[1] Chapter 18.)

Table 7 Additional cases of NMSC in the US population (whites only)

Scenario	Ozone status 2050 (%)	Existing population (million)	Cohort born 2030–74 (million)
1	+2	−0.13	−4.5
2	−1	0.45	1
3	−4	1	7
4	−10	2	92
5	−35	6	151
6	−50	12	160

Adapted from Environmental Protection Agency,[1] Chapter 18.

effect of ozone depletion will appear more quickly in statistics for melanoma than for NMSC. A short latent period between exposure and onset of melanoma is also supported by studies showing a relationship with sunspot activity.[56–58]

Figure 7 is based on age-adjusted incidence rates for melanoma in Connecticut, which date back to 1935. Cyclical changes in sunspot activity are associated with a temporary rise in melanoma incidence, which begins when sunspot activity is at its peak, and persists for 3–5 years.[57] However, it is not altogether clear how this relationship is mediated since years of maximal sunspot activity are associated with decreased rather than increased ultraviolet radiation at ground level.

Whether the effect of ozone depletion on melanoma incidence can be documented will depend on the severity of the ozone loss and the level of public

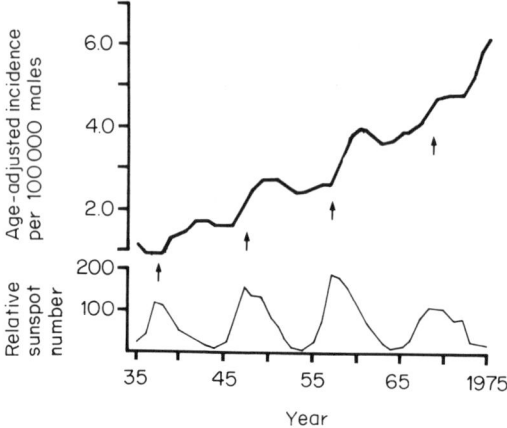

Figure 7 Annual age-adjusted incidence rates for malignant melanoma in Connecticut per 100 000 males and sunspot activity (arrows indicate peaks). (Reproduced from Houghton et al.[57])

Table 8 Additional melanoma deaths in the US population (whites only)

Scenario	Ozone status 2050 (%)	Existing population	Cohort born 2030–74
1	+2	−100	−10 000
2	−1	1 000	−2 000
3	−4	2 000	13 000
4	−10	4 000	112 000
5	−35	10 000	164 000
6	−50	18 000	170 000

Adapted from Environmental Protection Agency,[1] Chapter 18.

awareness. My prediction is that the high mortality rate of CMM will persuade most people to modify their sunbathing habits, and a decline in package holidays to Mediterranean countries will result. If people do not modify their behaviour, the EPA have estimated the additional melanoma deaths associated with each of their six scenarios. Table 8 shows an extra 18 000 melanoma deaths in the existing population using the worst case scenario, and for a population born between 2030 and 2074 an extra 170 000 melanoma deaths.

Cataracts

Cataract formation is the other medical condition that shows a latitude-dependent incidence. Most of the radiant energy reaching the lens is in the UVA range, with wavelengths below 300 nm contributing less than 1% of the total. On the other hand, ultraviolet light at 290 nm is 250 times more potent at inducing cataract formation experimentally than light at 320 nm.[59] Senile cataracts are immensely common. They usually contain both cortical and nuclear opacities, and their frequency increases with age. In the USA over 600 000 operations are performed annually. Many of these cataracts are probably an inevitable accompaniment of the ageing process rather than being specifically ultraviolet related. However, the incidence of nuclear brunescent cataract does show an inverse relationship with latitude.[60] It is known, for example, that the percentage of cataracts of the nuclear brunescent variety increases at lower latitude, being 9% in Rochester, New York (42°N), 20% in Tampa, Florida (28°N) and 43% in Manila (15°N).

In addition, the prevalence of cataracts in rural aborigines in Australia is higher than for white Australians and is strongly ultraviolet dependent.[61] Table 9 documents prevalence for three different age ranges in five different zones and demonstrates a consistent increase with age, and with increasing ultraviolet exposure. In the oldest age group the percentage of the population effected is less than 14% in zone 1, and almost 30% in zones 3, 4 and 5.

Table 9 Variations of cataract prevalence with UV dose in rural Aborigines

Ultraviolet zone	Age		
	0–39	40–59	60+
1	0	1.7	13.6
2	0	2.6	24.2
3	0.1	3.7	29.5
4	0.1	3.8	30.5
5	0.2	5.1	29.8

For populations at higher latitude who spend less time outdoors, the relationship is less obvious, but still discernible. In 1971–72 the National Health and Nutrition Examination Survey carried out 10 000 eye examinations in 35 geographic areas of the USA. Analysis of these data revealed a significant difference in the prevalence of cataracts in locations with larger amounts of sunlight.[62] Subsequently, the EPA used these data to construct a linear relationship between cataract risk and UVB flux (Figure 8). It is of interest that the percentage increase in prevalence decreases with age, so that a 30% increase in UVB flux will increase cataract prevalence by 17% at age 50 and by only 11% at age 70.

The EPA went on to compute the additional number of senile cataracts that would be seen in the existing US population using their six emission scenarios for chlorine- and bromine-bearing substances.[1] Estimates in Table 10 range from

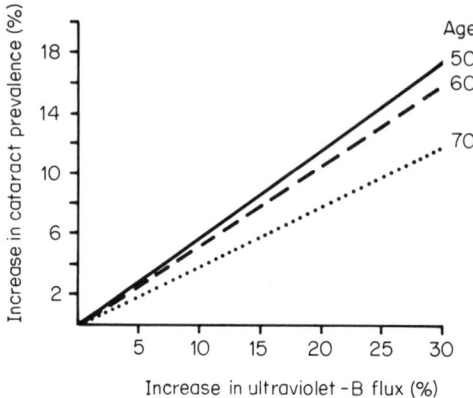

Figure 8 Estimated relationship between cataract risk and UVB flux (measured with an RB meter). (Source: Environmental Protection Agency.[1])

Table 10 Additional cases of senile cataract in the US population

Scenario	Ozone status 2050 (%)	Existing population	Cohort born 1986–2029
1	2	10 000	0
2	−1	103 000	49 000
3	−4	267 000	752 000
4	−10	614 000	4 312 000
5	−35	1 830 000	8 928 000
6	−50	3 239 000	10 361 000

Adapted from Environmental Protection Agency,[1] Chapter 18.

10 000 to 3 million additional cataracts using the best and worst case scenarios. People born in the USA over the next 40 years can expect up to 10 million extra cataracts, unless efforts to control CFC production are effective on a worldwide basis.

CONCLUSION

This paper has concentrated mainly on the direct health effects of stratospheric ozone depletion on human health, but damage to the ozone layer will not occur in isolation. Other processes will interact to destabilize atmospheric conditions and this will exert effects on the world community that could dwarf those from increased ultraviolet exposure (Figure 9). For example, trace gases responsible for ozone depletion also exert a powerful greenhouse effect. Loss of ozone from

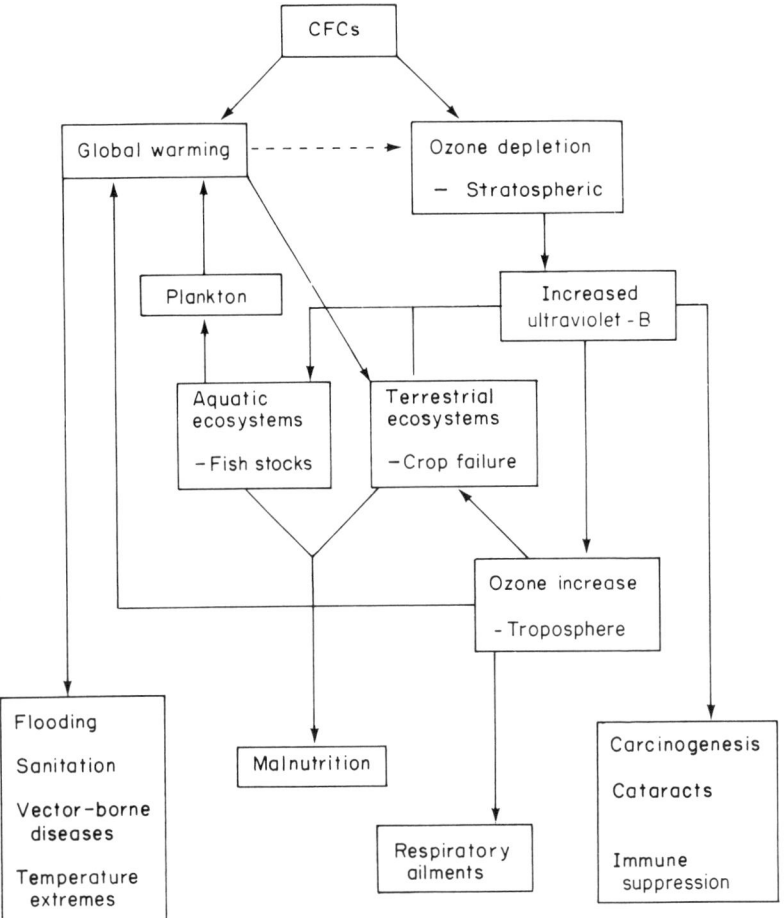

Figure 9   Mechanisms by which CFCs and other trace gases might exert effects on human health.

the stratosphere may increase the formation of tropospheric ozone, itself a greenhouse gas. Equally, the greenhouse effect may cool the stratosphere, thus promoting conditions suitable for further losses of stratospheric ozone. Finally, increased ultraviolet flux may affect plankton, reducing the ability of the oceans to absorb carbon dioxide and aggravating the problem of global warming. The prospects for human health are not good. The lack of sanitation from widespread flooding, the respiratory ailments associated with higher levels of tropospheric ozone, the spread of vector-borne infectious diseases that might follow greenhouse warming, and the malnutrition that could affect millions if crops fail are potentially catastrophic events whose gravity is matched only by their unpredictability.[63]

For my part, I have no doubt that the issue of stratospheric ozone represents the litmus test of man's ability to prevent the ultimate degradation of planet Earth. If this relatively simple problem cannot be solved, how can mankind hope to survive the vastly more complex problems that will arise in the centuries to come?

REFERENCES

1. Environmental Protection Agency, Assessing the risks of trace gases that can modify the atmosphere. US Environmental Protection Agency. Washington DC (1987).
2. J. van der Leun, Yearly review: UV carcinogenesis. *Photochem. Photobiol*, **39**, 861–868 (1984).
3. R. Russell Jones, Ozone depletion and cancer risk. *Lancet*, **ii**, 443–446 (1987).
4. R. Mackie and M. Rycroft, Health and the ozone layer (editorial). *Br. Med. J.*, **297**, 369–370 (1988).
5. M. Giannini, Suppression of pathogenesis in cutaneous leishmaniasis by UV irradiation. *Infect. Immunol.*, **51**, 838–843 (1986).
6. C. Spellman and R. Daynes, Modification of immunological potential by ultraviolet radiation. II. Generation of suppressor cells in short-term UV-irradiated mice. *Transplantation*, **24**, 120–126 (1977).
7. M. Fisher and M. Kripke, Systemic alteration induced in mice by ultraviolet light irradiation and its relationship to ultraviolet carcinogenesis. *Proc. Nat. Acad. Sci. USA*, **74**, 1688–1692 (1977).
8. P. Friedmann, Ultraviolet carcinogenesis in mice and men. *Br. J. Dermatol.*, **109**, 683–686 (1983).
9. R. Mackie, J. Elwood and J. Hawk, *Links between Exposure to Ultraviolet Radiation and Skin Cancer*. Report of the Royal College of Physicians, UK, (March 1987).
10. S. Wan, R. Anderson and J. Parrish, Analytical modeling for the optical properties of the skin with *in vitro* and *in vivo* applications. *Photochem. Photobiol.*, **34**, 493–499 (1981).
11. J. Norlund, Genetic basis of pigmentation and the disorders of pigmentation. In A. Ackerman, ed., *Pathology of Malignant Melanoma*. Masson Publishing, New York, pp 23–45 (1981).
12. J. Scotto and J. Fraumeni, Skin cancer (other than melanoma). In D. Schotterfeld and J. Fraumeni, eds, *Cancer Epidemiology and Prevention*. W. B. Saunders, Philadelphia, pp 996–1011 (1982).
13. J. Elwood, J. Lee and S. Walter, Relationship of Melanoma and other skin cancer morbidity to latitude and ultraviolet radiation in the United States and Canada. *Int. J. Epidemiol.*, **3**, 325–332 (1974).
14. D. Gordon and H. Silverstone, Worldwide epidemiology of premalignant and malignant cutaneous lesions. In R. Andrade, ed., *Cancer of the Skin*. W.B. Saunders, Philadelphia, pp 405–434 (1976).
15. F. Urbach, J. Epstein and P. Forbes, Ultraviolet carcinogenesis: experimental global and genetic aspects. In T. B. Fitzpatrick *et al.*, eds, *Sunlight and Man — Normal and Abnormal Photobiological Responses*. University of Tokyo Press, Tokyo, pp 259–283 (1974).

16. M. Kripke, F. Urbach and C. Witkop, Ultraviolet radiation carcinogenesis. In O. D. Laenum, ed., *Biology of Skin Cancer.* Biology of Human Cancer Workshops Report 15, pp 195–222 (1981).
17. J. Robbins, K. Kraemer, M. Lutzner *et al.*, Xeroderma pigmentosum: an inherited disease with sun-sensitivity, multiple cutaneous neoplasms, and abnormal DNA repair. *Ann. Int. Med.*, **80**, 221–248 (1974).
18. K. Kraemer, M. Lee and J. Scotto, Xeroderma pigmentosum: cutaneous ocular and neurological abnormalities in 830 published cases. *Arch. Dermatol.*, **123**, 241–250 (1987).
19. M. Kripke, Latency, histology, and antigenicity of tumours induced by ultraviolet light in three inbred mouse strains. *Cancer Res.*, **37**, 1395–1400 (1977).
20. J. Spikes, M. Kripke, R. Conner *et al.*, Time of appearance and histology of tumours induced in the dorsal skin of C3Hf mice by ultraviolet radiation from a mercury arc lamp. *Natl. Cancer Inst.*, **59**, 1637–1643 (1977).
21. L. Kligman and A. Kligman, Histogenesis and progression of ultraviolet light-induced tumours in hairless mice. *J. Natl. Cancer Inst.*, **67**, 1289–1297 (1981).
22. J. Streilen, The skin as an immune organ. In J. Parrish, ed., *The Effect of Ultraviolet Radiation on the Immune System.* Johnson & Johnson Baby Products, New Jersey, pp 21–34 (1983).
23. E. Smith and M. Brysk, Immunity and skin cancer. *S. Afr. Med. J.*, **74**, 44–46 (1981).
24. I. Penn, Depressed immunity and the development of cancer. *Clin. Exp. Immunol.*, **46**, 459–474 (1981).
25. I. Blohme and O. Larko, Premalignant and malignant skin lesions in renal transplant patients. *Transplantation*, **37**, 165–167 (1984).
26. A. Sheil, Cancer after transplantation. *World J. Surg.*, **10**, 389–396 (1986).
27. A. Gupta, C. Cardella and H. Haberman, Cutaneous malignant neoplasms in patients with renal transplants. *Arch. Dermatol.*, **122**, 1288–1293 (1986).
28. W. Clark, D. Elder and M. Van Horn, The biologic forms of malignant melanoma. *Hum. Pathol.*, **17**, 443–450 (1986).
29. V. McGovern, H. Shaw, G. Milton *et al.*, Is malignant melanoma arising in a Hutchinson's melanotic freckle a separate disease entity? *Histopathology*, **4**, 235–242 (1980).
30. K. Kraemer, M. Lee and J. Scotto, DNA repair protects against cutaneous and internal neoplasia — evidence from xeroderma pigmentosum. *Carcinogenesis*, **5**, 511–514 (1984).
31. J. Lee and A. Carter, Secular trends in mortality from malignant melanoma. *J. Natl. Cancer Inst.*, **45**, 91–97 (1970).
32. J. Elwood and J. Lee, Recent data on the epidemiology of malignant melanoma. *Sem. Oncol.*, **2**, 149–154 (1975).
33. O. Jensen and A. Bolander, Trends in malignant melanoma of the skin. *World Health Stat. Q.*, **33**, 2–26 (1980).
34. J. Lee and O. Strickland, Malignant melanoma: social status and outdoor work. *Br. J. Cancer*, **41**, 757–763 (1980).
35. J. Elwood, R. Gallagher, J. Davison *et al.*, Sunburn, suntan and the risk of cutaneous malignant melanoma: the Western Canada Melanoma Study. *Br. J. Cancer*, **53**, 543–549 (1985).
36. K. Cooke, D. Skegg and J. Fraser, Socioeconomic status, indoor and outdoor work, and malignant melanoma. *Int. J. Cancer*, **34**, 57–62 (1984).
37. J. Elwood, R. Gallagher, G. Hill *et al.*, Cutaneous melanoma in relation to intermittent and constant sun exposure: the Western Canada Melanoma Study. *Int. J. Cancer*, **35**, 421–433 (1985).

38. C. Holman, B. Armstrong and P. Heenan, Relationship of cutaneous malignant melanoma to individual sunlight exposure habits. *J. Nat. Cancer Inst.*, **76**, 403–414 (1986).
39. D. Elder, M. Greene, E. Bondi *et al.*, Acquired melanocytic naevi and melanoma: the dysplastic naevus syndrome. In A. Ackerman, ed., *Pathology of Malignant Melanoma*. Masson Publishing, New York, pp 185–215 (1981).
40. W. Epstein, J. Bystryn, R. Edelson *et al.*, Nonmelanoma skin cancer, melanomas, warts and viral oncogenesis. *J. Am. Acad. Dermatol.*, **5**, 960–970 (1984).
41. J. Dunn, E. Levin, G. Linden *et al.*, Skin cancer as a cause of death. *Calif. Med.*, **102**, 361–363 (1965).
42. National Academy of Science, *Causes and Effects of Stratospheric Ozone Radiation: An Update*. National Academy Press, Washington DC (1982).
43. R. Nixon, A. Dorevitch and R. Marks, Squamous cell carcinoma of the skin: accuracy of clinical diagnosis and outcome of follow-up in Australia. *Med. J. Aust.*, **144**, 235–238 (1986).
44. C. Holman and B. Armstrong, *Cancer Mortality Trends in Australia 1910–1979*. Cancer Council of Western Australia, Perth, pp 283–292 (1982).
45. National Cancer Institute, *1985 Annual Cancer Statistics Review presented to the National Cancer Advisory Board*, National Institute of Health, Bethesda, Maryland (1985).
46. E. Sondik, J. Young, J. Horm *et al.*, *1985 Annual Cancer Statistics Review*. National Cancer Institute, Bethesda, Maryland (1985).
47. G. Giles, R. Marks and P. Foley. Incidence of non-melanocytic skin cancer treated in Australia. *Br. Med. J.*, **296**, 13–17 (1988).
48. A. Green, Incidence and reporting of cutaneous malignant melanoma in Queensland. *Aust. J. Dermatol.*, **23**, 105–109 (1982).
49. J. van der Leun and F. Daniels, *Biologic Effects of Stratospheric Ozone Decrease: A Critical Review of Assessments*. Climatic Impact Assessment Programme. Monograph 5, Part 1. US Department of Transportation, Washington DC, pp 105–124 (1975).
50. R. Setlow, The wavelengths in sunlight effective in producing skin cancer: a theoretical analysis. *Proc. Natl. Acad. Sci. USA*, **69**, 3363–3366 (1974).
51. J. De Luisi and J. Harris, A determination of the absolute radiant energy of a Robertson Berger meter sunburn unit. *J. Atmos. Env.*, **17**, 751–758 (1983).
52. D. Berger, The sunburning ultraviolet meter: design and performance. *Photochem. Photobiol.*, **24**, 587–593 (1976).
53. C. Cole, P. Forbes and R. Davies, An action spectrum for UV photocarcinogenesis. *Photochem. Photobiol.*, **43**, 275–284 (1986).
54. T. Fears and J. Scotto, Estimating increases in skin cancer morbidity due to increases in ultraviolet radiation exposure. *Cancer Invest.*, **1**, 119–126 (1983).
55. J. Scotto, T. Fears and J. Fraumeni, *Incidence of Nonmelanoma Skin Cancer in the United States*. National Cancer Institute, US Dept of Health and Human Services, NIH 82-2433 (1981).
56. A. J. Serdlow, Incidence of malignant melanoma of the skin in England and Wales and its relationship to sunshine. *Br. Med. J.*, **2**, 1324–1327 (1979).
57. A. Houghton, E. Munster and M. Viola, Increased incidence of malignant melanoma after peaks of sunspot activity. *Lancet*, **i**, 759–760 (1978).
58. D. Wigle, Malignant melanoma of skin and sunspot activity. *Lancet*, **ii**, 38 (1978).
59. D. Pitts and A. Cullen, Determination of infrared radiation levels for acute ocular cataractogenesis. *Albrecht von Graefes Arch. Clin. Exp. Ophthalmol.*, **217**, 285–297 (1981).

60. S. Zigman, M. Datiles and E. Torczynski, Sunlight and human cataracts. *Invest. Ophthalmol. Visual Sci.*, **18**, 462–467 (1979).
61. F. Hollows and D. Moran, Cataract: the ultraviolet risk factor. *Lancet* , **ii**, 1249–1251 (1981).
62. R. Hiller, R. Sperduto and F. Ederer, Epidemiologic associations with cataract in the 1971–1972 national health and nutrition examination survey. *Am. J. Epidemiol.*, **118**, 239–249 (1983).
63. Editorial. Health in the greenhouse. *Lancet*, **i** , 819–820 (1989).

Discussion Period 5

Dr Everest (UK Centre for Economic and Environmental Development) Dr Russell Jones, how can you justify asking a country such as India, which has not signed the Montreal Protocol, to curtail what they see as essential use of CFCs for their development, to solve what appears to be essentially a Caucasian problem, multiplied and exacerbated by Western lifestyle and even colonialism?

Dr Russell Jones It is certainly true that black races are largely immune to ultraviolet-induced skin cancer. However, no-one is immune to the immunosuppressive effects of ultraviolet light, and black races will not be immune to the formation of cataracts. Also, I think even India is beginning to realize that some of the so-called natural disasters that afflict their sub-continent are not entirely natural, and that some of them are man-made. I think that one might be dealing with a sympathetic audience once this has been pointed out.

Dr Scorer (Imperial College) I am worried by Dr Russell Jones's final remark that if we cannot solve this problem then we don't have much hope of solving others. One cannot disagree with that, but on the other hand this problem is one that has arisen because of the increase in human activity on the Earth's surface. If we still had only the population of the year 1900 this issue would not even have become a problem.

Dr Russell Jones I cannot deny that industrial pollution is related to the number of people on the surface of the planet. The reason that I see this as a litmus test is because the chemicals that produce stratospheric ozone loss are not an irreplaceable part of our society and many of the uses are relatively trivial. Society has developed over countless generations without the benefit of these chemicals and will doubtless survive perfectly well without them. We know what causes this problem and we know what the solution is. It is not even very

---

Ozone Depletion: Health and Environmental Consequences
Edited by R. Russell Jones and T. Wigley
© 1989 John Wiley & Sons Ltd

expensive to implement the solution. We need to get this particular problem right because there are much more difficult problems around the corner.

Dr Morton We have been told that total column ozone has fallen over northern latitudes, but has this been paralleled by an increase in ultraviolet flux at the surface of the Earth? Are we receiving a greater amount of ultraviolet radiation than previously?

Dr Russell Jones The Ozone Trend Review Panel showed that there were modest losses over northern latitudes between 1970 and 1987. That should increase the amount of ultraviolet reaching the surface of the Earth, but measurements by RB meters in North America show that this is not in fact the case. That does not necessarily mean that there is not more ultraviolet light coming through. What it could mean is that the meters are located in areas of high local atmospheric pollution. So what is needed is a network of meters that measure ultraviolet irradiance directly, located in non-urban areas. When that is established, data can be obtained to see what the chronological trends are.

Dr van der Leun I just want to support that last statement. There is no established global network for UVB irradiance, wavelength by wavelength, and this is something that is definitely required. The network needs to be established now so that we can obtain baseline information for the future.

Don Cusack (United States Department of Agriculture) In the UNEP Conference held in 1977 and before that in the White House Report, which came out in 1975, we made a strong plea for unweighted ultraviolet irradiance data. Essentially, a complete solar cycle of data has been missed. The RB network is now facing serious funding problems. The Manufacturing Chemists Association and its Fluorocarbon Panel have approached a number of government agencies to carry on some of that funding. We now have to decide whether the programme should be continued until a worldwide network is set up, or we will go through a phase when there is no data at all for a couple of years. I also understand that the TOMS Satellite Programme is within a few years of becoming obsolete. The average life of the satellites I understand is something like 12 years and they have been up about 10 years. So we are not only facing serious concern at the troposphere level but also in the stratosphere. And again I would like to pose a challenge to this conference. What steps are we going to take to address these very important issues? We have this problem of trying to reconcile the Ozone Trend Panel data with the measurements of ultraviolet at ground level. How can one make a valid biological assessment that is going to be useful *vis-à-vis* the Montreal Protocol if the basic data are missing?

DISCUSSION PERIOD 5

Dr van der Leun   I would like to respond. The RB meter network has served a very valuable purpose in establishing a latitudinal gradient for UVB exposure. It is not a very suitable indicator for determining long-term trends in UVB irradiation. Moreover, these meters have been designed to imitate one action spectrum whereas, with ozone depletion, we are dealing with many biological effects, which in some cases have very different action spectra. I agree with you wholeheartedly that we do need to establish a monitoring network to obtain reliable long-term data, but we need instrumentation that is far more appropriate for this type of measurement and in the correct location. However, let us not throw away old shoes before we have new ones.

Bob Watson   It was suggested that the TOMS instrument is now 12 years old and will soon expire. I agree with that. However, we should also look forward to the fact that we have three ways to measure ozone, two of them routinely. There is the Dobson network. This has its limitations but calibration is improving through the efforts of the WMO. Secondly, NASA has an operational SBUV instrument that measures ozone as well as incoming solar radiation globally. As far as the TOMS instrument is concerned, NASA is working very closely with the Soviet Government and we hope within 18 months to have an American instrument aboard a Soviet satellite. There are also plans to fly a similar instrument on a Japanese satellite in 1993/1994, the Adios satellite. So while this current Nimbus 7 TOMS instrument will almost certainly die within the next 18 months or so, there are plans to fly instrumentation aboard a Soviet satellite, a Japanese satellite and possibly even another American satellite.

Part 7
Political Aspects

17
Alternatives to CFCs

C. E. TANE
Imperial Chemical Industries PLC

INDUSTRY'S OBJECTIVE

First of all, I should like to thank the organizers of this conference for the opportunity to express the views of industry and to summarize the intensive activity that industry has undertaken as its response to the issue of CFCs and ozone. I shall set out the positive progress that has so far been achieved in a number of areas and go on to describe the challenges that lie ahead and which industry must face before we can be sure of complete success. I hope that I shall be able to demonstrate that, by and large, all of industry — and by this I mean not just the manufacturers of CFCs, such as ICI, but also all the many downstream companies whose products currently depend on CFCs — is committed to the task of finding alternatives to CFCs and is rising to the challenge.

Lest there be any doubt, let me make it clear that the objective I refer to is, quite simply, the eventual elimination of CFC usage from all current applications. ICI, along with a number of other major CFC manufacturers, has publicly accepted that this is the target and committed itself to a series of initiatives aimed at making this possible. We have made it clear that we believe the objective can be achieved only when new chemicals have been fully developed for a number of very important applications, and that this process will require time.

APPLICATION OF CFCs

To understand the reasoning behind this assertion, let us briefly look at the principal applications of the major CFCs. These are as propellants for aerosols,

---

Ozone Depletion: Health and Environmental Consequences
Edited by R. Russell Jones and T. Wigley
© 1989 John Wiley & Sons Ltd

as a blowing agent for foam-blown plastics used in packaging and insulation, as refrigerants in air-conditioning circuits and refrigerating systems, and cleansing agents for electronic circuitry. Within each of these divisions are subcategories.

The solutions, and the timing of CFC replacement, differ between the different subcategories and it can be very misleading to generalize about the ease of introducing alternatives to CFCs in particular categories. For example, although hydrocarbon (butane/propane) aerosol propellants can substitute for CFCs in personal care aerosols, e.g. deodorants and hairsprays, they are not at all suitable for medical aerosols, such as inhalers for asthma sufferers, or for many industrial aerosol products. We cannot say that CFCs can be substituted in all aerosol applications until solutions have been found for the most demanding area of use.

It is clear that CFCs are used in a wide variety of applications, some of which — such as the refrigeration of food — are in anybody's terms vital to our modern lifestyles. In a few cases, alternative technologies may be available which eliminate the need for any chemical 'active ingredient'; in most cases, however, a substitute chemical compound needs to be developed. It can be easily demonstrated that, although the list of potential chemicals is theoretically huge, the choice in reality is extremely narrow, and, as a result, substitutes that have limitations of one kind or another may need to be employed.

PROPERTIES OF CFCs

To understand why this is, we need to look at the reasons why CFCs became so widely used before their potential to deplete ozone was properly understood.

As many of you will know, when CFCs were first invented they were hailed as wonder chemicals. The low toxicity of CFCs has meant that they have been able to be used in a wide range of applications from aerosols to domestic refrigerators and insulating foam, without posing a health risk to the consumer. In the case of solvent cleaning, the development of CFC 113 took place specifically as part of a move away from more toxic substances being used as solvents in industry. The non-flammability of CFCs has been another important contributor to the wide availability of products containing them, and to safety in the factories in which those products are made.

Because CFCs are very simple molecules, they can be produced by relatively easy chemical processes. This means that they can be made on a wide scale (over 20 countries worldwide manufacture CFCs) and produced cheaply. If they had not been so affordable, the products that use them would have been less affordable and therefore many of the goods now taken for granted might have been available only to the wealthy.

The chemical stability of CFCs has always been one of their major assets, and, as we now know, is their major drawback from the environmental point of view. Nevertheless, stability has been crucial to many of the purposes for which CFCs have been used. Domestic refrigerators, in which the refrigerant must circulate continuously around the system for ten years and more, could not operate without a stable fluid that does not corrode or otherwise react with the materials of construction.

CRITERIA FOR ALTERNATIVES TO CFCs

In view of the properties that made CFCs so useful, it is easy to identify the criteria that any substitute should ideally meet. Clearly, environmental stability, in particular potential to affect stratospheric ozone, is of paramount importance.

That the substitute should be of low toxicity, ideally as low as the CFCs, goes without saying. It also needs to be capable of being manufactured easily, firstly to keep extra costs to users and ultimate consumers within manageable proportions, and secondly so that the manufacture of the substitutes can spread across the globe, just as it did for CFCs, thus offering all regions of the world the security of supply of the alternatives that will enable more and more countries to eliminate CFC use.

It is very important that any substitute should have very similar physical properties to those of the CFC it is intended to replace, i.e. that it should as far as possible be a 'drop-in' replacement. The physical properties that matter vary from application to application, but the effect of significant differences in physical characteristics is considerable. Even a small difference will require the user industries to invest in major expensive retooling or reformulation. The greater the difference, the bigger the investment required and the more difficult it becomes for industry to afford it. In some cases, the cost or the difficulty will be so great that they cannot be accommodated by some companies, who might therefore be forced to close, with consequent job losses. In the case of aerosol manufacturers, many will be able to formulate away from non-flammable CFCs to flammable hydrocarbon propellants at a cost they can afford; but there are also aerosol fillers who, due to geographical location, will not be able to obtain approval to store and handle flammable propellants in their factories.

Unfortunately, it is a fact that no two chemicals ever have exactly the same physical properties and there are therefore no true drop-ins, only approximations to them. The consequence of this is that the investment required of industry to make and use alternatives to CFCs will be immense. The cost of building sufficient capacity for manufacture of alternatives will in itself be a drain on capital within the chemical industry for the next 10–20 years; but the cost of retooling so that downstream industries can use the substitutes properly and safely will be much more. Estimates of the cost of this replacement process generally forecast expenditure of several billion pounds.

EXISTING ALTERNATIVES TO CFCs

Let us now look at those broad areas of application of CFCs for which alternatives already exist. As I said, no alternative can fully replicate the physical properties of CFCs and each of the existing alternatives has some disadvantages in comparison with the CFC it is intended to replace.

The British Aerosol Manufacturers Association has recommended to its members that they eliminate the use of CFC aerosol propellants by the end of 1989. The propellants that will be used to replace CFCs will be hydrocarbons (butane/propane mixtures) and possibly also dimethyl ether (DME). The most important drawback here is the flammability of these substitutes.

As a result of the flammability, significant modifications at considerable expense must be made to aerosol filling factories in order to handle hydrocarbon propellants; and there are additional requirements for storage of large quantities of such aerosols. Formulation difficulties arise in some areas because of the different solubility characteristics of the hydrocarbon propellants, but these will be overcome in due course. Finally, hydrocarbon propellant emissions can contribute to another environmental problem — that of low-level smog formation.

In the area of polystyrene foam blowing for food packaging, two alternative chemicals can be used. HCFC 22 will be discussed later; the other candidate, pentane, is another flammable hydrocarbon, and the same considerations apply as for aerosols.

In some types of polyurethane foam, the CFC blowing agent can be partially replaced by using water to generate carbon dioxide. This technology, recently developed by ICI Polyurethanes as well as other companies, leads to significant reductions in the insulation value of the finished foam and thus can only be used in applications in which either the insulation capability is not of importance or other conditions can be varied to compensate. Thus the water/carbon dioxide technology is far from a complete answer.

HCFC ALTERNATIVES

The applicability of HCFCs as alternatives to CFCs is probably the most controversial topic of discussion within industry today. I use the term here to include both the partially halogenated CFCs such as HCFC 22 and also those chemicals, such as methyl chloroform, that are not strictly CFCs, but can also contribute chlorine to the stratosphere and thus have small ozone depletion potentials.

HCFCs that can be substituted for CFCs are already available on the market and the substitution process can be started immediately. Because the HCFC alternatives are very different physically from the CFCs they are intended to replace, the substitution process will be long and very costly.

Other new HCFCs are being developed and chemical producers such as ICI will need to take decisions soon as to whether to invest in full-scale manufacture. In both cases, very large amounts of investment are required, and the companies concerned, both users and manufacturers, will want to be as certain as possible that their investment will be a good one.

In reality the messages that industry receives about HCFCs are far from conducive to taking major investment decisions. Views expressed by legislators, environmental pressure groups and government agencies vary widely, from seeing HCFCs as 'part of the solution, not part of the problem' on the one hand, to bringing forward legislation attempting to reduce or ban use of HCFCs on the other.

To tell industry that HCFCs are available as part of the solution, but only for the next 5 years, is the worst of all worlds. Although accounting convention may write investment off over that period, all companies in my experience expect their investments to produce returns over considerably longer periods, and are unlikely to pursue an investment with such a short lifespan. In this climate of uncertainty, decisions whether to go with HCFCs or to wait for other solutions as yet unknown to present themselves are simply not being taken .

I will not attempt to address the scientific issues relating to HCFCs, since other delegates are far better qualified to do that than I. As a representative of industry, however, I would stress that a clear message on HCFCs is vital, otherwise we risk taking decisions that may result in wasted investments and ultimately loss of jobs. Needless to say, it would also undermine the credibility of the environmental lobby in the eyes of many parts of industry, which I suggest would be a significant retrograde step.

HFC 134a

I would like to move on now to talk about one of the most positive areas of progress achieved by industry, and that is the work done to develop production and use of HFC 134a, the 'ozone-benign' substitute for CFC 12. As you may know, ICI has recently announced plans to invest in two plants to make HFC 134a, the first in the UK and the second in America. The first plant is expected to come into full production early in 1991. DuPont have also announced a US plant to come on stream at about the same time. This timetable will cut the period normally required to develop a new chemical in half and has been possible only because new methods of project management have been developed.

A number of challenges still lie ahead before widespread use of HFC 134a can be assured and CFC 12 can be phased out of use in many of its major applications. Apart from the need for adequate toxicological data to be developed, most of the outstanding difficulties are technical problems relating to application of the product in practice. It is a sign of our determination to achieve the earliest possible launch of the new product that we have announced investment plans before toxicology test results are fully available, and before we are completely certain that customers will be able to adapt their technology to use it. The technical problems relate to the fact that, although HFC 134a is about the nearest to a drop-in for CFC 12 that can be achieved, it is by no means an exact replica and therefore the hardware needs to be adapted to use it.

Lubrication of refrigeration systems remains the biggest hurdle. The lubricant formerly used with CFC 12 is incompatible with HFC 134a and therefore a new lubricant must be developed. ICI's Lubricant Group is confident that a suitable material can be found, but it will be much more difficult to find a lubricant compatible with *both* HFC 134a and CFC 12. The implication of this is that in refrigeration and air conditioning systems that require regular topping up with refrigerant, HFC 134a cannot be used if CFC 12 and its lubricant are already in the system.

Likewise, CFC 12 and HFC 134a can have very different effects on some of the materials of construction of refrigeration equipment. Certain materials that may be used in the manufacture of, for example, gaskets and O-rings swell in contact with CFC 12, whilst others shrink; and the design of the system and the choice of materials takes this into account. These materials often react differently in contact with HFC 134a. Again, this implies that HFC 134a cannot simply be added to a refrigeration system designed for CFC 12; new materials may need to be selected and thus new equipment specifically designed to be compatible with 134a has to be developed.

In turn, this implies that supplies of CFC 12 will continue to be required to provide top-up quantities for existing CFC-based systems until they reach the end of their economic life, possibly up to 20 years later. If this compatibility problem cannot be overcome in some way, the need for CFC 12 could continue well into the next century, unless legislation is brought forward to mandate the replacement of CFC-based systems with new systems using HFC 134a. This would of course require significant new investment, and might strain the output capacity of refrigeration equipment manufacturers.

One further determinant of the speed with which HFC 134a can replace CFC 12 will be the availability of 134a. Although ICI and DuPont's planned investments are a positive start, the volumes involved fall a long way short of the quantities required fully to replace CFC 12, production of which runs into hundreds of thousands of tonnes per year. There are risks associated with investment in alternatives to CFCs that do not normally apply to chemical investments, particularly in view of the compressed timetable. Although ICI has clearly demonstrated that it is prepared to take these additional risks, it is likely that many of the smaller CFC manufacturers worldwide will want to wait until technical problems have been overcome and the market is clearly defined before they invest themselves. It is very difficult to predict how long it will take before sufficient 134a capacity is available to replace CFC 12, but it is likely to be some considerable time.

CONCLUSION

In conclusion, ICI and the vast majority of its customers accept that the ultimate objective must be the elimination of CFC production and use. Considerable progress has already been made by ICI, its co-producers and its customers, particularly in the development of manufacture and use of HFC 134a. Development of substitutes for CFC 11 is also well advanced.

Technical problems must still be overcome before the substitutes to CFCs can be used in all areas of current CFC applications, and the difficulties involved in overcoming these problems and the time required to do so should not be underestimated. Ideal solutions to the replacement of some of the smaller volume grades of CFC, such as 113 and 115, have still to be found, unless it is accepted that chlorine containing chemicals such as methyl chloroform and HCFC 22 are acceptable substitutes. In view of the billions of pounds of investment required to complete the industrial restructuring necessary, economic considerations may also prove a stumbling block.

Nevertheless, I hope it is clear as a result of today's presentation that industry has taken on the challenge of eliminating the use of CFCs and is fully determined to achieve it. Time does not permit detailed reference to industry's achievements to date in the area of recovery, recycling and emission reduction, but here too significant progress is being made. If solutions to the problem of CFCs and ozone can be found, they will be found; all that is required is time to develop them.

18
Environmental Imperatives

JONATHON PORRITT
Director of Friends of the Earth, UK

I am delighted to have the opportunity to address such a prestigious gathering on behalf of Friends of the Earth, who, through Robin Russell Jones, the Chairman of our Pollution Advisory Committee, first conceived the idea of this Conference.

I would like to use this opportunity to pay tribute to an extraordinary coalition of scientists, environmentalists, consumers and international agencies, whose work has done so much to accelerate the introduction of measures to protect the ozone layer. The degree of interdependence between the different participants in this coalition has been crucially important in the UK, where environmental campaigning on the ozone layer was at a low ebb between 1978 and 1985. Only when the scientific evidence began to harden, particularly through Joe Farman's work with the British Antarctic Survey, did the environmental campaigning harden proportionately. Friends of the Earth was then able to use a consumer-based campaign to bring pressure to bear firstly on the aerosol industry, a campaign which progressively involved the Consumers Association and bodies like the Women's Institute, creating a unique if largely uncoordinated alliance. The very welcome cooperation of Friends of the Earth and the Consumers' Association in co-sponsoring this Conference provides powerful confirmation of the effectiveness of such an alliance.

While this alliance was coming together in the UK, a few far-sighted governments were working with and through the United Nations Environment Programme (UNEP) to give birth to the Montreal Protocol, which many of us in the Environment Movement consider to be UNEP's single most important achievement since it came into being. Such an achievement is all the more

---

Ozone Depletion: Health and Environmental Consequences
Edited by R. Russell Jones and T. Wigley
© 1989 John Wiley & Sons Ltd

remarkable when you consider both the extent to which UNEP has been starved of proper funding and the extent to which Peter Usher's work has been starved of funding within UNEP itself.

To that list of 'good guys' (scientists, environmentalists, consumers and UNEP), we may now, apparently, add the UK Government. It would be fair to say that this did not appear to us a realistic eventuality when we first conceived the idea of this Conference more than a year ago. It is a genuine surprise to all environmentalists that the Government which, rightly or wrongly, has earned Britain the epithet of 'The Dirty Man of Europe' should have been transformed so rapidly into the ozone layer's Green Fairy Godmother!

We unhesitatingly welcome this transformation, and pay tribute to those in the Department of the Environment who have been beavering away to make it possible, and to the determination of Lord Caithness and Mrs Bottomley to make this a real priority. The recently announced initiative of a major international conference to encourage other countries to adopt the Montreal Protocol is particularly welcome.

However, before we all get carried away with a warm afterglow of euphoria at the conversion of one environmental sinner, we must first analyse a little more carefully what all this amounts to, and expose in particular the claims made on behalf of the so-called 'voluntary approach' to a rather more rigorous analysis than we have heard at this Conference so far.

It will have not escaped your attention that Mrs Bottomley made several eulogistic references to the voluntary approach in her opening address to the Conference. The UK Government is in the envious position of being able to claim international credit for meeting the Montreal Protocol's target of a 50% reduction in CFC consumption *a full ten years* ahead of the target date. It is this breakthrough that has allowed them to campaign so actively (and so welcomely) for an 85% target when the Protocol is reviewed.

As you have already heard, the 50% reduction in the UK will be achieved almost exclusively because the aerosol industry (which has up until now been responsible for about 60% of the use of CFCs in the UK) will be phasing out its use of CFCs as propellants by the end of 1989. As it happens, the Government had little, if anything, to do with that decision, which was taken by the aerosol industry under intense pressure from Friends of the Earth and consumer organizations.

That pressure came in two distinct but complementary varieties. Firstly, through the publication of our pamphlet *The Aerosol Connection* (a detailed listing of all aerosols that are not using CFCs as propellants), coupled with as much publicity and consciousness-raising as we could generate at the time. Secondly, when this 'softly-softly' approach failed to elicit anything more than vaguely hostile rebuffs of our scientific evidence, we started preparing an outright boycott of the best-selling CFC-based products of leading companies in the UK, to be coordinated throughout our network of 250 local groups. The aerosol industry's decision to get out of CFCs by 1989 was taken just three days before the boycott campaign was launched.

This was a sensible and pragmatic decision by the industry. Their own advisers had kept them right up to date with the debate about the hole in the ozone layer over the Antarctic, but many were at that time still not convinced as to the *direct* linkage between CFCs and ozone depletion. It was actually the fear of consumers turning against aerosols *per se* (not just CFC-based aerosols) that tipped them over the edge, rather than some rational decision based on objective science. Consumer awareness is, in fact, a somewhat rudimentary weapon, and the industry accurately read the signs of what was happening. Once Prince Charles declared that he had banned *all* aerosols from his Household, the Royal imprimatur was well and truly on the wall.

I should add that this one, almost throw-away, line from Prince Charles did more to allow the UK to meet the Montreal Protocol target than *anything* the Government has done. The Government's new-found environmental fervour is clearly in danger of obscuring the facts. It would, after all, be extremely misleading simply to disregard the months of assiduous lobbying by the UK Government, primarily at ICI's behest, for the EEC to press for a mere freeze, or at best a mere 20% reduction, in CFCs within the framework of the Montreal Protocol.

What then should we conclude from the campaign to date against the use of CFCs in aerosols? First, the Government should be allowed to take absolutely no retrospective credit whatsoever for what has been achieved, which is essentially the consequence of good science, successful environmental campaigning, and the welcome, if somewhat belated, pragmatism of the aerosol industry.

Secondly, the voluntary approach should be seen as a flight of ideological fantasy, of minimum intellectual value in this country and absolutely valueless in countries where the possibility of such pressure being brought to bear on industries simply does not exist. It is one of those pleasing paradoxes of the Government's new turquoise tendency that it is actually *dependent* on environmental pressure groups to do the job it should be doing itself. A voluntary approach would seem to mean nothing more than leaving it to voluntary organizations, as happened on the issue of lead-free petrol, which the Government is now promoting via the services of CLEAR, the self-same pressure group they once spent years trying to do down. The same is now happening with CFCs and the ozone layer, with the Government entirely dependent on Friends of the Earth and consumers to put pressure on the industry to phase out CFCs.

Thirdly, as Stanley Clinton Davis diplomatically but so succinctly spelled out in his talk yesterday, the voluntary approach is no substitute for a proper programme of mandatory regulation by government. There are many sectors of CFC use that are far less vulnerable to consumer pressure than aerosols. This has been clearly reflected in the tactics Friends of the Earth has had to use in its campaign. After the aerosol industry, we turned our attention to those using foam packaging materials blown with CFCs, particularly the supermarkets and fast-foods outlets. The latter responded with admirable promptitude to the idea of a national advertising campaign by Friends of the Earth based on the theme of fast-food packaging as the Ozone Take Away Industry, and it is good news that today all major fast-food companies are now using alternatives to CFCs 11 and 12.

Beyond this, our ability to use direct consumer pressure is clearly constrained. The construction industry, for instance, is a major user of CFCs for insulation materials and air-conditioning parts, but is at least one degree removed from direct contact with the consumer. In this instance, we have therefore chosen to commission a major piece of research on CFC uses in the construction industry and the potential for alternatives and substitutes. We will be using this throughout 1989 as our major lobbying tool with all the major construction companies within the UK.

Lastly, we have already had discussions with the representatives of the refrigeration industry, who clearly face a very different set of problems. Since the debate is now so much further advanced than it was 2 years ago, the industry is already very anxious to discuss alternatives to CFCs among themselves, with a view to finding solutions through constructive dialogue.

The lesson here is a simple one. Even in a country with a relatively powerful environment and consumer movement, the greater the gap between the product and the consumer, and the more complex the process of substitution, the more compelling becomes the case for direct regulation, with governments setting absolutely clear phase-out dates for all CFC uses in order to accelerate the process of innovation. There are many examples of this approach in other countries, particularly Canada, the USA, Sweden, Norway and Denmark.

The UK Government claims that legislating for change does not work, and cites as their 'evidence' for this the fact that CFC use in the USA has steadily *increased* since the compulsory ban on CFCs in aerosols in 1978. The rather obvious fact that CFC use over that period would have increased far faster had they still been in use in aerosols merely demonstrates how low a government will sometimes stoop to rationalize its own ideological incoherence.

Mandatory phase-out periods should go hand in hand with the provision of reasonable incentives to promote the fullest possible retrieval, recycling and safe disposal methods of and for CFCs. As yet, the importance of this continues to elude our Government, as demonstrated in this revealing answer from Lord James Douglas-Hamilton to a Parliamentary Question on 26 October 1988 from Mr Hume Robinson about assistance for local authorities in Scotland over the safe disposal of unwanted refrigerators and other equipment containing CFCs. The answer read as follows:

> The vaporization of CFCs does not present any significant risk to human health and no advice to local authorities is considered necessary. There is at present no

environmentally safe disposal method that local authorities could use, but my Right Honourable Friend the Secretary of State for the Environment, through his Department, actively pursues means to reduce emissions with user industries.

The problems of retrieval and disposal are primarily those of the developed world. Third World countries face a very different burden as a result of phasing out CFCs. Environmentalists believe that tangible assistance should be made available to Third World countries to cushion the increased financial burden of having to adapt to the new generation of CFC substitutes. It serves little purpose berating these countries for wishing to expand their limited CFC use when the existing threat to the ozone layer has come about almost exclusively because of CFC usage by the developed world. It is therefore unacceptable for any western government to bring pressure to bear on Third World countries *unless* they are prepared to protect them in some way from the inevitably higher costs of CFC substitutes. Any financial assistance should be provided outside the existing aid arrangements between the developed world and the Third World.

The question of CFC substitutes is rightly coming to dominate the scientific debate. Mr Tane of ICI complains in Chapter 17 about the uncertainty of the position taken here by environmentalists in that we have refused to give outright support to the introduction of partially halogenated CFCs, such as HCFC 22, which has a far lower ozone depletion potential than CFCs 11 and 12. I am grateful to Mr Tane for his solicitude about the potential threat to the credibility of Friends of the Earth, but it seems to us to be very reasonable to argue that *all* chemicals that contribute chlorine to the stratosphere are undesirable, however small their ozone depletion potential, though it is true that there is a strictly limited transitional role in weaning certain industries from the more damaging CFCs. Massive investment in HCFC 22 would, therefore, be extremely unwise.

It is not just a question of the ozone depletion potential of substitutes; the contribution of such substitutes to the greenhouse effect is of crucial importance. Keith Shine (Chapter 7) said that if HFC 134a was introduced as a total substitute for CFC 12 (allowing for modest growth of around 3%), this would contribute about 20% of the global warming effect that would have come from the equivalent use of CFC 12. That still constitutes a highly significant contribution to global warming and one that we should countenance with the utmost reluctance.

This may of course lead to the uncomfortable conclusion that in some circumstances there is *no* substitute that is both ozone-friendly and neutral in global warming terms. If, at the end of the day, when all potential substitutes have been rigorously examined for their ozone depletion potential and greenhouse effect impact, there is no new wonder chemical that meets suitably rigorous environmental criteria, then perhaps we will finally realise that we should in fact be talking a rather different language. Perhaps at that stage our thoughts will turn to a society geared more to genuine need and sustainable development, rather than to the financial dictates of industry; to the question of how to design buildings that don't need air conditioning, or transport systems that do not depend on covering the planet with air-conditioned limousines with their own built-in cocktail fridges. Frightening though it must sound to some of the scientists in the audience today, it may just materialize that there is *no* technical fix to saving the planet from global warming.

At moments during this Conference I have therefore felt a bit like a hostage who has been held by terrorists for several months with a gun to his head. On realizing that the gun may actually backfire, the terrorists suddenly relent in their determination to kill me — and then expect to be thanked profusely for this act of outstanding charity!

Industry and government have been holding a metaphorical gun to the atmosphere for many years, unwittingly to begin with, but latterly quite knowingly. Environmentalists have no intention of falling over themselves with gratitude for the lowering of this gun, especially as we are far from convinced that a whole new arsenal of so-called 'environmentally benign' guns aren't about to be put in its place.

We do not necessarily believe that either industry or the UK Government has learned the real lessons of CFCs and the ozone layer. What is one to make, for instance, of ICI's ambivalence in calling on the one hand for a complete phase-out of CFCs, but refusing on the other to rule out the possibility of *increased* exports of CFCs in volumes at least as great as will be saved by cuts in the UK market as we succeed in meeting the Montreal Protocol target? This loophole in the Protocol is a very serious problem; commercial profitability and global responsibility are uneasy bedfellows at the best of times. Would it be too much to suppose that ICI's undoubted concurrence in the Government voluntary approach will prevent them from taking advantage of the loophole?

Efforts currently afoot to widen the ozone and global warming agendas (by including other chemicals in the Montreal Protocol, for instance) do not give one grounds for much confidence. Methyl chloroform is one such chemical. It may indeed have a very short atmospheric lifetime and a low ozone depletion potential, but because of the scale of its use it is already contributing more to ozone depletion than CFCs 114 and 115 or any of the halons already involved in the Montreal Protocol.

The US Environmental Protection Agency has calculated that chlorine-containing chemicals *not* covered by the Protocol will account for 40% of projected growth in stratospheric chlorine by 2075. Methyl chloroform will contribute 80% of this 40%.

ICI's response to this was disturbingly predictable, referring to the EPA report as 'somewhat alarmist', and as containing 'wholly unrealistic projections'. Have we not heard that kind of thing before, not so very long ago, with CFCs? Should we not fear another gun being raised to our heads? The chances of finding appropriate and *timely* solutions to the greenhouse effect are remote indeed if industry digs in behind its customary defensive barriers, which then have to be dismantled brick by brick, as they have been with CFCs.

That is why we should be leaving this Conference acknowledging both what has been achieved (through the alliance of good science, uncompromising environmentalist and consumer action, and properly resourced international watchdogs) *and* what still needs to be done. The sad truth is that whilst industry and some governments are keen to be seen patting their respective environmental watchdogs on the head, they are still working to their own agenda, which is shaped more by vested commercial interest and competitive nationalism than by any nascent concern for the well-being of Planet Earth.

To leave the maintenance of that planet to such fatuous fictions as the voluntary approach, to the good intentions of industry, or indeed to the inevitably limited resources of environmental and consumer organizations is simply not good enough. Hence our recognition that we still have a very long way to go before we can lower our defences on the ozone layer, let alone on the whole question of the greenhouse effect.

19
Ozone Depletion: Consumer Choice

RACHEL WATERHOUSE CBE
Consumers' Association

Ten years ago it might have been surprising to find the Consumers' Association co-sponsoring a conference concerned with environmental pollution. Are we just bandwaggoning? I hope not. We are, in fact, representing and responding to our members by taking up their own expressed anxieties about health and safety. We have over a million members subscribing to our magazines and as a consumer organization we need to keep close to them and to know what their concerns are — what they want us to keep them informed about. We achieve this in great measure through regular and *ad hoc* surveys of our readers and their opinions.

Last July, for example, we carried out a postal survey of a sample of our members and asked them about their environmental concerns and, if they were concerned or worried, what they had done about it.

The survey began by asking respondents to rank as 'a very serious problem', a 'fairly serious problem', a 'not very serious problem' or 'no problem at all', eight major issues on the political agenda.

Table 1 shows the rankings under a 'very serious problem' and a 'fairly serious problem'. Three issues came out clearly at the top of the poll: drugs; law and order; and environmental pollution. If we take the very serious problems alone, drugs and environmental pollution are equally ahead of law and order. If we take 'very serious' and 'fairly serious' together, law and order draws ahead, but these three issues are very closely bunched.

Ozone Depletion: Health and Environmental Consequences
Edited by R. Russell Jones and T. Wigley
Published 1989 by John Wiley & Sons Ltd © Consumers' Association

252 OZONE DEPLETION: HEALTH AND ENVIRONMENTAL CONSEQUENCES

Table 1

	Very serious problem (%)	Fairly serious problem (%)
Environmental pollution	47	38
Drugs	47	39
Law and order	43	44
Nuclear waste	38	32
The National Health Service	35	38
Unemployment	31	48
Quality of education	30	42
Teenage drinking	22	43

Source: Consumers' Association Survey, July 1988, Base 1117.

Our next question asked specifically which of 20 environmental issues the respondents were most concerned about.

The top six, as shown in Table 2, were river and sea pollution (43/42%), aerosols and the ozone layer (41/37%), nuclear waste (41/33%)followed by insecticides, fertilizers and other chemicals used in agriculture (35/39%), dirty beaches and bathing water (35/44%) and passive smoking (35/30%). Again, if we take 'very serious' and 'fairly serious' together, aerosols and the ozone layer comes second in the table.

Table 2

	Very concerned (%)	Fairly concerned (%)
River and sea pollution	43	42
Aerosols and the ozone layer	41	37
Nuclear waste	41	33
Dirty beaches and bathing water	35	44
Insecticides, fertilizers, etc.	35	39
Passive smoking	35	30

Source: Consumers' Association Survey, July 1988, Base 1117.

Respondents were then asked which environmental issue they regarded as a potential threat to their own health, with the results shown in Table 3. 'Very serious threat', with 'very serious' and 'fairly serious' together, were nuclear waste (35/26%), aerosols and the ozone layer (33/37%), passive smoking (26/32%), asbestos (25/24%), radiation (25/23%), insecticides, fertilisers, etc. (24/41%) and river and sea pollution (23/39%).

These three slides illustrate that consumers show a surprisingly high concern about aerosols and the ozone layer. The problem, of course, impacts on

Table 3

	Very serious problem (%)	Fairly serious problem (%)
Nuclear waste	35	26
Aerosols and the ozone layer	33	37
Passive smoking	26	32
Asbestos	25	24
Radiation	25	23
Insecticides, fertilizers, etc.	24	41
River and sea pollution	23	39

Source: Consumers' Association Survey, July 1988, Base 1117.

consumers directly and without distinction and on a world scale in a way that few others do. In the global context, destruction of the ozone layer may threaten world food production, and the accumulation of CFCs (along with other gases we produce, like carbon dioxide), by increasing the greenhouse effect, may contribute to the melting of the polar ice caps.

As was said in the closing statement at Montreal 'Humanity is conducting an enormous, unintended, globally pervasive experiment second only to global nuclear war'. From the consumer's point of view it has this one great advantage: decisions do not all have to be left to politicians. Nuclear armaments divide nations; a natural disaster can bring them together. It is indeed now politicians who are trying to catch up with and climb aboard the environmental express.

For individual consumers there are two further spurs to action. Concern about the global environmental problem is sustained by the more immediate personal threat of increased skin cancers as the sun's ultraviolet penetrates the thinner or non-existent ozone layer. This personal concern is reflected in Table 3. But, secondly, consumers feel that they can make a personal contribution to improving the situation.

We have come a long way in a surprisingly short time. When we published our first article on aerosols in *Which?* magazine in June 1980, the state of the art was such that we felt we had to lay some emphasis on a statement that 'it has still not been proved conclusively that CFCs are damaging the atmosphere'. We also had to report that in the 6 years since 1974 when the ozone theory was first put forward there had been a continuing increase in the use of aerosols with no information about the propellants used. That article, therefore, included a list of aerosols that normally used CFCs and of a small number of manufacturers and retailers who claimed not to use them in particular products. We concluded the article, 9 years ago, by suggesting that consumers could help by buying non-CFC aerosols.

We also said that 'we do think that manufacturers should be made to say which propellant they've used, so you could avoid those which contain CFCs'.

Unlike the Netherlands, where compulsory labelling for CFCs 11 and 12 was introduced in 1979 (a decision that threw up a minefield of technical problems), the UK has not made labelling mandatory. However, such information is essential to consumer choice. If those looking for ozone-friendly aerosols cannot distinguish them from others, the alternative is to give up using aerosols altogether. In our July survey this year we asked what our concerned members had done specifically to protect themselves or the environment from the possible effects of pollution. Only 44% of the sample had done anything, but of that 44%, as Table 4 shows, 72% had stopped using aerosols altogether.

Table 4

	%
Stopped using aerosols	72
Reduced/stopped use of insecticides, fertilizers or other chemicals used in agriculture	31
Eating organic food	25
Joined/contributed to environmental organisation	24
Buying unleaded petrol	20
Stopped smoking	18

Source: Consumers' Association Survey, July 1988, Base 493.

Many aerosol cans are still not labelled with the type of propellant they contain, but in February this year, eight aerosol manufacturers (with a claimed 65% of the UK aerosol market) announced that they would be phasing out the use of fully halogenated CFCs as propellants in aerosols by the end of 1989. Even if several manufacturers, representing this claimed two thirds of the UK aerosol market, are phasing out CFCs by the end of 1989, this phase-out will not be total, and of course there may still be imports. Many of these manufacturers are already using ozone-friendly labels, which is clearly useful for the consumer. We very much welcome the UK aerosol industry's change of heart, but we now need to go much further than this, by ensuring that all products that contain CFCs are clearly labelled as such. Clear and reliable information of this kind is essential if consumers are going to be able to play their part in protecting the environment.

We have already done what we can to publicize the facts in an article in *Which?* in July this year. Here, we were able to give a much longer and fuller list of aerosols under own-brand and manufacturers' names in the three categories of toiletries, perfumes and household products with proposed phase-out dates.

The tremendous build-up of consumer resistance to aerosols in the UK over the last 8 years has, I have no doubt, contributed to the early voluntary phase-out by manufacturers.

Moreover, this voluntary action has enabled the British Government to claim that, by the end of 1989, the UK will be 10 years ahead of the Montreal Protocol, the main provision of which is for a 50% reduction in 1986 levels of CFC consumption by the end of the century — a decision in which percentages favour the apparent virtues of those who start from a base of previous inaction.

In Europe, where we work closely with consumer testing organizations in other EC countries, the Bureau of European Consumer Unions (BEUC) joined with the European Environmental Bureau last year in a campaign to reduce CFCs in consumer products. A primary target was aerosol reformulation. After negotiations with the Federation of European Aerosol Associations in July of this year, the Federation directed all its members to sign voluntary agreements which foresee a reduction of at least 90% in the use of fully halogenated CFCs in aerosols within 2 years.

Of course, this is not the only source of CFCs in the atmosphere, but recent evidence to the House of Lords Select Committee on the European Communities estimated that 60% of UK consumption of fully halogenated CFCs was in aerosols. Fortunately, for the vast majority of aerosols, there are alternative propellants or alternative products — although we are aware of the continuing need for CFCs in some medical products. As our survey showed, consumers not fully informed by labelling of the propellants seem to have taken the other route and have chosen to give up aerosols entirely.

It may not be so easy for consumers directly to influence use of CFCs in other processes. However, for our July 1988 *Which?* report we asked various fast-food and supermarket chains about the boxes and trays they use for burgers, for example, or for meat trays, which use CFC 11 as a blowing or foaming agent. McDonalds and Casey Jones said they had already phased out CFC-containing packaging in the UK. Burger King use cardboard boxes; Spud-U-Like told us they have several packaging suppliers — their main one does not use CFCs. Wimpy said then that they were 'considering' changing to boxes blown with less harmful CFC 22, and we know that this has now been virtually accomplished.

The five supermarket chains we asked — Asda, Gateway, Safeway, Sainsbury and Tesco — all said that their packaging suppliers plan to stop using CFCs by the end of the year or as soon as possible.

It would help if we all as consumers asked at our local supermarkets and fast-food chains if their packaging was CFC free.

When we turn to domestic refrigerators, since CFCs are unlikely to leak during use, the problem becomes one of waste disposal. If the refrigerant could be recycled or safely destroyed, that would be a satisfactory solution, but it seems to be too difficult and costly to be practical at the moment. Even so, it is important that further research should be carried out, and we are pleased to hear that pioneering work is being done in Cologne, in Scandinavia and the Netherlands, to see what schemes might be appropriate. There seems to be only a limited contribution individual consumers can currently make to solve this problem, but given some practical arrangements there is every hope that they will respond. One possibility might be arrangements for old fridges to be traded in when new ones are bought — rather like car batteries at the moment. And there is surely a role for a pioneering local authority to blaze the trail here by starting up a scheme for recycling fridges, if only on an experimental basis. In addition to domestic refrigerators, of course, increasingly widespread use of commercial freezer units, for preparing, storing and distributing frozen food, compounds the problem. Here again some workable recycling arrangements are likely to be a more practical solution than consumers refusing to buy frozen foods.

Air-conditioning systems that use CFCs are more difficult, as here leakage does occur, especially in vehicles or when an industrial plant is serviced inefficiently. Better equipment and a high standard of servicing to prevent leakages, together with rapid development of safe alternatives for CFCs, would seem to be the best way forward. Consumers need to be reassured, however, that whatever substitutes are developed they are no threat to health and that they make no contribution to the greenhouse effect while protecting the ozone layer.

The Montreal Protocol has now been signed by the UK. All EC member states will have signed by the end of the year. We welcome the British Government's new-found enthusiasm for protection of the ozone layer and the international conference they are to call, as announced last week. International agreement about reduction in manufacturing and use of CFCs is essential if products unacceptable in developed countries are not soon to be found surfacing in Third World industries. I was attending an International Organization of Consumer Unions (IOCU) meeting in Madrid in September 1987 when consumers from some 55 countries were waiting to hear whether the agreement in Montreal had been signed. Developing countries have long felt threatened by our use of CFCs. They were overjoyed at the news of the Protocol. They will campaign fiercely to prevent any further threat to the world that is theirs and ours.

The title of this chapter is Consumer Choice. I have sought to show how individual consumers, given information and choice, can make a positive contribution to changing patterns in the marketplace. They are deeply concerned about environmental issues and rate highly the independent, consumer-oriented and practical advice we are able to provide, so that they may make their own contribution to the protection of the world they live in.

20
Priorities for Research

D. A. WARRILOW
Department of the Environment, UK

The subject of stratospheric ozone and its depletion has developed extremely rapidly, in line with the remarkable events that have taken place within the 1980s in the stratosphere, particularly over Antarctica. Indeed, so rapid has been the change that many speakers and participants have been intimately connected with the subject over this period, some carrying out important new research. The Antarctic research campaign of 1987 yielded large amounts of new data and led to greatly increased understanding of the processes that cause severe ozone depletion over the Antarctic continent. It was confirmed that chlorofluorocarbons (CFCs) were indeed responsible for the ozone depletion, in conjunction with the special meteorological conditions existing there.

There has been considerable activity in the last few months. In October 1988 the UK Stratospheric Ozone Review Group published its second report, which had an important impact on UK policy. Also in that month a major step was taken towards coordination and cooperation of stratospheric ozone research within Europe, and the first in a series of meetings leading to a review of the Montreal Protocol on Substances that Deplete the Ozone Layer was held.

Stratospheric problems must also be seen in the context of global atmospheric change and the greenhouse effect, chlorofluorocarbons being strong greenhouse gases. This year has seen this reflected in discussions at a number of international conferences including the Toronto Conference on the Changing Atmosphere in June and more recently in Hamburg. More significantly, the Inter Governmental

---

Ozone Depletion: Health and Environmental Consequences
Edited by R. Russell Jones and T. Wigley
Published 1989 by John Wiley & Sons Ltd © Crown copyright 1989. Reproduced by permission of the Controller of Her Majesty's Stationery Office

Panel on Climate Change, which met in Geneva under the aegis of UNEP/WMO, has set in motion a programme of assessment of our scientific understanding of the implications of increasing concentrations of greenhouse gases.

One might think that following all this activity, and given that major international moves are now afoot to deal with ozone depletion, we now fully understand the problem. It is true that major advances have been made, however there remain significant gaps in understanding, particularly of the detailed chemical processes and the significant shortcomings in our ability to model the depletions that have been observed.

Therefore the purpose of this talk is to

(1) Review the areas of uncertainty.
(2) Discuss what measures might be taken to improve our understanding.
(3) Discuss what research activities are being planned at the present time.

I shall confine myself to speaking mainly about the current and planned activities in scientific research relating to the stratosphere from a mainly UK and European perspective, but will briefly discuss the greenhouse component and work on effects and alternative materials.

AREAS OF UNCERTAINTY

Despite the considerable effort of recent years, significant areas of uncertainty remain. Our knowledge of concentrations of trace constituents and their chemical reactions in the stratosphere is limited. The importance of heterogeneous chemistry in the Antarctic stratosphere is well known, but its role in the Arctic and Equatorial stratosphere is not clear, where nevertheless it is known that stratospheric clouds exist. The possibility of heterogeneous reactions on aerosols is still speculative. The details of many reaction mechanisms and kinetics are unknown, especially at low temperatures and pressures. As recently as 2 years ago, dynamic theories for the appearance of the Antarctic ozone hole were seriously considered. Even now, although we have enough evidence that chemical processes are mainly responsible, the role of dynamic processes in transport of trace contents needs to be clarified. Finally, modelling studies are unable to simulate present ozone trends, underestimating them by about a factor of 2 in some cases. Models do not yet incorporate heterogeneous chemistry and so are unable to simulate the Antarctic hole. Use predominantly of one- and two-dimensional models means that transport processes are inadequately represented.

RECOMMENDATIONS FOR FUTURE RESEARCH

The Stratospheric Ozone Review Group report,[1] published in October 1988, made a number of recommendations for future research in the UK, although

much of this has already been carried out in the USA and Europe. As the Review Group points out, there remain fundamental scientific uncertainties concerning our understanding of and ability to predict anthropogenic impacts on stratospheric ozone.
The need for future work can be identified under three broad headings.

Observation

This is the baseline activity. The scientific community needs to continue to monitor stratospheric ozone both from the ground and from satellites. In addition, it would be desirable to measure the concentrations of a host of trace constituents such as nitrogen dioxide, nitric acid, nitric oxide, water, hydrogen peroxide, methane, carbon monoxide, CFC 12 ($CF_2 Cl_2$), CFC 11 ($CFCl_3$), hydrogen chloride, ClO, $ClONO_2$, N_2O_5 and hydrogen fluoride at a larger number of locations. The vertical concentration structure of these constituents might also be usefully monitored. Two types of observation are important: long-term observations with good spatial coverage which give the needed overall context of changes, temporal and spatial; and short-term, detailed observations, perhaps at a higher spatial and temporal resolution, which provide information from which chemical processes can be deduced. The latter may require the use of aircraft measurements. In some cases it is the only way in which some chemical species can be observed.

Such an expansion in observational requirements requires development of instrumentation such as ground-based microwave radiometers, development of ground-based LIDAR for measurement of polar stratospheric clouds, and development of techniques for observing from aircraft or balloons. It is generally assumed that ozone depletion will lead to increased levels of short-wavelength UVB (290–320 nm) at the surface. In fact, there is virtually no information on UVB trends at the surface and it would seem to be necessary to consider the need for spectral observations of UVB to confirm or otherwise this effect on ozone depletion. At the same time it would be useful to observe the levels of the longer-wavelength UVA at the surface. Some estimate of ultraviolet trends can of course be deduced from radiation changes based on ozone totals.

There are a number of difficulties to be overcome in long-term measurement of ultraviolet radiation, not least one of instrument calibration. It may, in fact, take a decade or more before any real trend would be observed, because of the large variance in the ultraviolet signal and because of instrumentation difficulties.

Chemical reaction laboratory studies

Recent studies have revealed a number of novel chemical features, which need to be investigated thoroughly. For example, such studies should include the chemistry and photochemistry of the ClO dimer, and this is already being carried out at Harwell for DOE; the kinetics and products of the heterogeneous reactions of the reservoir molecules ($ClONO_2$, N_2O_5, HOCl) with water/ice, hydrogen chloride and sulphate aerosols; improvement of kinetic data from bromine reactions; attention to reactions at low temperatures; and formation of stratospheric clouds. In addition to furthering understanding of observed chemical concentrations, these studies are very important for providing information for chemical modelling studies.

Theoretical and modelling studies

Modelling studies help us to understand chemical processes occurring in the stratosphere and can provide us with forecasts of trace gas concentrations in the atmosphere, including those of stratospheric ozone. These models are indispensable in alerting us to potential ozone depletion and providing us with quantitative estimates of the impact of reduced emissions of ozone-depleting substances. Therefore, in order to know how much emission reduction is required to stop or reverse ozone depletion, it is vitally important that the models are developed further. A hierarchy of models of varying chemical and dynamic complexity are required. Sound empirical and theoretical descriptions of constituent transport need to be developed. Reliable quantitative estimates of the global distributions of fluxes from the troposphere to the stratosphere and vice versa, including rainout, are very important for long-term predictions. Those are not available to us at the moment. Specific studies need to look at Antarctic and Arctic processes as well as more general extra-polar ozone depletion. Heterogeneous chemistry requires to be included in future modelling work. Reduction in ozone changes the direct heating of the atmosphere by solar radiation and can therefore potentially affect the general circulation. Work in this area should be carried out using GCM models.

COOPERATIVE PROGRAMMES

It is clear that although individual groups can and do make an enormous contribution to the science of the subject, they nevertheless often achieve their goals while working on larger projects and programmes. Atmospheric research is of such a scale that it often requires large resources in both finance and manpower and a high degree of inter-institute and international cooperation. This is certainly true of field programmes, but may become true of modelling programmes as these grow in complexity. Therefore it should be no surprise to find that the present trend in stratospheric research is towards collaborative and coordinated programmes, to deal with the scale of the task and to make best use of available resources.

Firstly, SORG[1] recommended that a greater degree of cooperation and coordination of stratospheric research should be encouraged in the UK. Currently, research is funded by a number of bodies, among them the research councils SERC and NERC, the Meteorological Office (which has a significant stratospheric programme), and DOE (which has a continuing programme of support for laboratory and modelling studies as well as field programmes where such work helps in providing a base for developing policy and advising Ministers). In addition, the National Physical Laboratory has been active in instrument development. The Review Group has played some part in focusing research direction, but controls no funding. They recommend that a national coordinating committee be set up, which will have the ability to establish and fund a coordinated programme. Moves to follow this recommendation are at an early stage. Already studies on types and availability of aircraft for research have been made by some members of the Review Group.

Secondly, significant progress has been made towards coordinating stratospheric ozone research in Europe. Over the past year, following a UK initiative at an EC/EFTA Ministers meeting, programme managers and some scientists from EC/EFTA countries have met several times to discuss the nature of a coordinated programme and how it might be set up. Over the year, the plans have evolved from one of a loosely coordinated programme with a very light administrative structure comprising a committee of scientists who might meet a few times each year, to a more formal idea of a Task Force with a small permanent staff and a scientific advisory panel.

In October 1988 in the Hague a meeting of European programme managers and scientists discussed the Task Force structure and agreed to bring it within the EC/EFTA structure on cooperative atmosphere research, known as COST611. As it now stands, the Task Force will comprise a small permanently staffed unit of about two scientists which will be located somewhere in the UK. In addition, a scientific panel will provide the unit with scientific advice. The Task Force will review the work currently being undertaken in Europe, draw up a plan of possible collaborative research with particular subprojects, encourage participation in the programme, and encourage the shared use of facilities, etc. This project depends on the voluntary support of scientists in EC/EFTA countries and is something of an experiment, but which it is hoped will lead to greater cooperation between European countries in this area of science.

A cooperative project is being set up to look at the Arctic stratosphere in some detail in the spring of 1989. In this case, the work is being carried out jointly by the USA, the UK (mainly through the Meteorological Office) and Norway. It is already known that the Arctic stratosphere exhibits perturbed chemistry, but that ozone depletion is nowhere as extensive or as intense as in the Antarctic, because of the different meteorological conditions prevalent there. The aim will be to carry out a detailed observational study of the dynamical processes and chemical concentration using extra ground-based equipment, two aircraft and satellites. A further Antarctic campaign is being planned for 1990.

SCIENTIFIC ASSESSMENT

The pace of research and public response continues at a high level, and already the Montreal Protocol is being reassessed even before it comes into operation. The large body of new results makes that reassessment both timely and necessary, and the assessment process will be able to include results from the Arctic campaign.

The assessment process may itself initiate further work. For example, one important area in the assessment, which requires careful consideration, is the definition of Ozone Depletion Potential (ODP). ODPs for various CFCs and halons were provided by modelling studies and are not unique numbers. This area is particularly important as CFC alternatives are often partially chlorinated hydrocarbons (HCFCs) and have some, albeit small, ODP. The picture is complicated by the growth in the concentration of halons (CFC-like substances that contain bromine) since their ODPs increase with the amount of chlorine in the atmosphere. Additionally, and perhaps more importantly, improved estimates of the greenhouse warming potential of CFCs are required.

RESEARCH IN OTHER AREAS

I shall briefly mention here other areas of research work, which are clearly not insignificant.

When considering future research into the impact that changing UV levels might have on biological and materials receptors, a parallel can be drawn with the integrated approach taken by DOE to the impact of acid deposition. This includes a long-term monitoring programme to give the necessary baseline data, research into damage mechanisms on both laboratory and field scales, and some attempt to quantify the scale of damage done. As well as assessing the impact of increased ultraviolet radiation, other environmental stresses (both natural and anthropogenic) might need to be taken into account, such as increased low-level ozone, acid deposition and climate change.

In considering biological impacts, further work is needed to assess the likely risk of increased melanoma and non-melanoma, cataracts, interference with immunosuppression reactions, and increased sensitivity to disease in humans. In addition, the effects of ultraviolet light on natural ecosystems needs to be studied to answer questions about their sensitivity to changing environmental stress.

Man-made polymers are known to be affected by UVB radiation changing their colour, structural strength and characteristics, and the importance of this needs to be assessed.

It is important to remember that a significant area of research is one that seeks to deal with the problem at source. Production of alternative substances requires a considerable research and development effort — developing new substances, testing for toxicology, stability, etc., and development of production processes.

Clearly much work can be done to develop alternative technologies such as alternative means of producing sprays, electrical cleaning, making refrigeration equipment and so on.

CONCLUSIONS

Stratospheric research has developed rapidly in the last few years as a result of ozone depletion. Significant increases in understanding of atmospheric chemistry have taken place and these have been important in guiding the international and national responses to the problem. Nevertheless, areas of uncertainty remain which require further effort. In particular, improved observations of chemical concentrations are required, along with laboratory and modelling studies of chemical reactions, which must now include the so-called heterogeneous chemistry. Modelling studies need to be improved to give better indications of future ozone changes with different CFC control scenarios.

The scale of the task is such that most progress is likely to be made through cooperative and coordinated programmes. SORG has recommended that such coordination between UK funding bodies should be encouraged. Discussions on cooperation and coordination within Europe are well under way, with the formation of an EC/EFTA Task Force likely in the near future. An observational campaign will take place in the Arctic in the Spring of 1989, involving groups from the USA, the UK and Norway.

Much research needs to be done on impacts of ultraviolet increases on humans, ecosystems and materials. In the short term, a desk study of available information would help to clarify the extent of knowledge and clarify the areas that need to be studied.

ACKNOWLEDGEMENTS

The author would like to thank John Pyle, Geoff Jenkins and Bob Wilson for their helpful comments on this chapter. The views expressed are those of the author and are not necessarily those of the Department of the Environment.

REFERENCES

1. SORG, *Stratospheric Ozone 1988*, United Kingdom Stratospheric Ozone Review Group, 2nd Report, HMSO, London (1988).

Part 8
Conclusion

21
Concluding Remarks

LORD ZUCKERMAN

The close attention with which the discussions of the past 2 days have been followed is a measure of the widespread concern that exists about a major and new environmental problem, as well as a tribute both to those who worked so hard to bring this symposium about — and here I would mention in particular Dr Robin Russell Jones — and to the organizations by which it has been sponsored. In the few minutes that I have in which to add some concluding remarks, I can focus on only a few general points.

First, it is useful to remind ourselves that while public concern about the degradation of the global environment is a relatively recent phenomenon, warning lights about possible dangers have in fact been flashing for decades. What was early on called the problem of the conservation of nature had become a matter for serious concern well before the turn of this century. Our own Fauna Protection Society was founded as long ago as 1903 under the title of the Preservation of the Wild Fauna of the Empire — an empire that then covered half the globe. After the Second World War, international concern about the profligate way in which both renewable and non-renewable resources were being exploited became so great that in 1949 the young United Nations Organization convened an international conference to make the facts known. This meeting was followed in 1954 by the first international conference to consider the unrestrained growth of the world's population. Environmental pollution — a consequence both of population growth and increasing demand — had by then also become a major national and international worry.

In 1972 a major conference that the UN convened in Stockholm to consider the problems of the 'human environment' became the prelude to the establishment of

---

Ozone Depletion: Health and Environmental Consequences
Edited by R. Russell Jones and T. Wigley
© 1989 John Wiley & Sons Ltd

a permanent international organization that is responsible for coordinating action designed to prevent deleterious environmental changes. But, well before this, many industrialized countries had started to take action. It is 20 years now since our own Department of the Environment was established. The Royal Commission on Environmental Pollution, in whose launching I am proud to say that I was able to play a part, was set up at about the same time. I mention these things lest we forget that a considerable body of experience now exists to which one can refer when considering the practicalities of action designed to prevent or correct environmental damage.

The second point that I would make is that it is not at all surprising that different countries give different weight to the problems of the environment, with the consequence that international organizations such as the UN and the EEC have to struggle — frequently in vain — to try to make member states adopt or abide by common environmental standards. The reason why this is so is obvious. Repairing past environmental devastation and preventing further degradation costs money. Some countries can afford more than others. Some cannot afford anything. But in the end it does not matter where the money comes from, whether from government out of the public purse, from the polluter as a component of his costs of production, or from the ordinary citizen in the form of increased taxation or higher prices. Willy nilly, this means that environmental problems become fitted into different scales of priorities according to which countries allocate the totality of resources they command.

The third general point that I would like to make is that by global environmental degradation we mean the cumulative and transnational effect of the deleterious changes that may be occurring anywhere in the world. What is of great concern today is that many countries regard as their own business, matters that may have worldwide environmental consequences, many of which are with us now. Others have yet to reveal their effects in a sufficiently unequivocal way for effective international pressure to be brought to bear on the country or countries concerned. This is never easily done. Acid rain is a good example. This became an international issue well over 20 years ago, when Scandinavian governments started protesting that the damage to their forests and lakes was being caused by emissions into the atmosphere of sulphur and nitrogen oxides from the stacks of power and other industrial plants in the UK and the Ruhr. First we argued whether acid rain was a reality; then about the damage that it was alleged to cause; and finally about the question of national guilt. As a result, it is only in the last few years that a concerted programme has been launched to modify the offending stacks by fitting them with scrubbers. The basic reason for this long-drawn-out affair was the fact that to do anything about it meant the expenditure of large sums of money.

The lesson that can be drawn from this story is that, even when the relation of cause and effect can be firmly established in the environmental field, it does not follow that remedial action will necessarily be taken straight away. Action depends on a variety of considerations, most of which in the end add up to the question of cost and to the position the problem is accorded in the list of other national priorities. The ease with which a deleterious environmental change can be dealt with is equally important.

In the case of the greenhouse effect, we face a problem where we do not know all the factors that are involved, or which of those that we can already clearly identify are the most significant, or whether dealing with any single factor — for example, replacing fossil-fuelled power plants with nuclear stations — will not result in other environmental hazards; and sometimes whether it is politically possible to do anything meaningful to help slow down the secular rise in average global temperature, for example by preventing the destruction of the world's tree cover in Third World countries. Before touching on the ozone problem, there are two other points that I wish to make.

First, while it is always tempting, it can also be shortsighted to keep pointing an accusing finger at manufacturing industry as the cause of most of our environmental ills. We should not forget that the rise in average standards of living, in the extension of the expectation of life, and in the reduction of the burden of labour that has taken place in many countries has been due to the growth of manufacturing industry. The public not only accepts all the benefits that industry has provided, but goes on demanding more. Environmental degradation is part of the price that we have had to pay for our greed. Another of its consequences is the political unrest that derives from the uneven way the benefits of industrial progress are shared by the world's population.

In general, industry has become increasingly aware of its own responsibility to reduce the undesirable consequences of some industrial processes. More and more it is having to face the problem of internalizing the costs of preventing and repairing environmental damage; costs that before were 'externalized' by virtue of the implicit belief that it was the 'government's' responsibility to do whatever was called for. Today, governments are increasingly becoming environmental wardens. They can prohibit the manufacture and sale of hazardous products such as toxic agrochemicals, for example. They can demand strict toxicological tests before licensing the sale of new drugs. The laws of liability can be changed. And the principle of 'the polluter pays' can be imposed wherever this is possible. The extent and the rate at which social costs — of which the cost of protecting the environment is only one — can be internalized, that is to say transferred from the state to the producer and the consumer, is a matter for political decision. Our own Government's privatization programme implies a belief that the market mechanism will reveal the extent to which the public is prepared to pay for the protection of the environment. Not surprisingly, privatization, particularly of water and power, has become a highly contentious political matter.

What I have been saying will have indicated to you the context in which I view the ozone problem. It is only relatively recently that we have become aware that there is a problem. There is now practically no controversy about its nature, scale and cause. This Symposium has spelt out what the depletion of the ozone layer means for all of us, and to the future of the global environment. Relative to other major environmental problems it is easy to see what needs to be done, and there has been no argument on this point. I was therefore somewhat bewildered when I heard Dr Porritt say that the Friends of the Earth had drawn blood in so far as governments had taken action to deal with CFCs. Blood can be intoxicating. Governments had started to act on scientific advice about the ozone layer without waiting to be urged into action by pressure groups that had no special source of scientific information about the problem. It is ten years since the USA decided to ban the use of CFCs as propellants in sprays. There is international agreement now about the need to reduce the scale of their use. Our own Government has called for even speedier action than was agreed at the recent Montreal meeting. Industry has responded by searching for alternatives to CFCs, both in their use as propellants and in refrigeration plants and packaging. We have been told in this Symposium about the probable scale of the demand there will be for alternatives, and about the vast sums — and risks — of the investment that manufacturing industry will be called upon to make. At first sight the burden of these investments will have to be carried by the chemical industry. In the end, however, as I have already implied, it is a burden that we shall all be sharing.

Mr Clinton Davis has told us about the difficulty of getting the European Community to act in a concerted way in the field of environmental protection. I have long felt that what is lacking is a standing international advisory commission of independent scientists charged to keep environmental problems under review, in an orderly and objective way, as they become identifiable, and to which governments and international organizations alike could refer for what I would call dispassionate advice. The commission that I have in mind would have the right to make public its own views about the priority that should be accorded specific environmental issues, and about the costs that would be involved. What is wanting now is some kind of scientific court that would be independent of all national and international organizations, and which would not be regarded as a pressure group, focusing first on one and then another issue according to fashion.

I fear, however, that what I have in mind is not likely to come about soon. For the time being we have to hope that what needs to be done about the ozone problem will be done, whatever the difficulties involved in getting governments to act together. We do not want to see the depletion of the ozone layer becoming an unmanageable problem. Mr Clinton Davis has told us that he once asked a politician friend of his whether the lack of general public interest in environmental problems was due to ignorance or apathy, and received the answer 'I do not know, nor do I care'. That cynical response is, alas, not surprising when one thinks of the day-to-day tangible problems with which politicians wrestle, as opposed to the long-term environmental dangers whose effects are outside the scope of day-to-day experience. The action that has already been taken to deal with CFCs, and the further action that lies in store, may help bring about a greater public understanding of the need to prevent all deleterious changes for which humans are responsible; changes that spell danger to man's well-being and to the biosphere as a whole.

Finally, it only remains for me to thank once again the sponsors and all those who worked so hard to make this Symposium possible.

Index

acid rain 65, 268
The Aerosol Connection 244
aerosol sprays
  alternatives to CFCs 236, 237, 238
  consumer pressures 244–5, 252–5
  phasing out CFCs 7, 244, 254–5
aerosols, sulphate 31, 60
air-conditioning systems 240, 246, 256
air temperatures, marine, global mean changes 87–8
aircraft
  measuring Antarctic atmosphere 54–5
  supersonic, ozone depleting effects 31, 64, 198
albedo changes 88
algal blooms 102
amplification factors
  biological (BAF) 164–5, 167, 216–18
  Mcdonald's 162
  optical (radiation) (OAF) 166–7, 216
animal studies
  photocarcinogenesis 162–7, 170–1
  photoimmunological effects 167–8
Antarctic ozone hole 22–3, 52–6
  future prospects 41, 138, 199
  greenhouse effect and 81–2
  impact on development of Montreal Protocol 121, 123
  key features 52–3
  mechanisms 30–1, 54–6, 62–3, 258
Arctic stratosphere
  cooperative research project (1989) 199–200, 261
  mechanisms of ozone depletion 22, 24–5, 61

see also Northern Hemisphere
atmosphere
  changing chemical composition 44–6
  energy transfer within 72–6
  future monitoring 259
  layers xi
atmospheric pollution
  ground-level UV levels and 181, 191
  international control 139
atmospheric window 75–6

basal cell carcinoma (BCC) 161–2, 210–11
  calculated increase in incidence 164, 167, 216–18
  at different latitudes 209–10
  in organ transplant recipients 212
  in xeroderma pigmentosum 148, 211
behavioural change, melanoma risk and 180, 182–3, 184, 220
biological amplification factor (BAF) 164–5, 167, 216–18
British Government, *see* UK Government
bromine xii, 17, 61
bromine monoxide (BrO) radical 31, 55
Bureau of European Consumer Unions (BEUC) 255

carbon dioxide
  faint Sun paradox and 106
  infrared absorption 73, 75, 79
  predicted ozone levels and 30, 37–9, 45, 47
  projected changes 33, 91, 92

carbon dioxide (contd.)
  reducing emissions 9, 69–70, 96
  role in global warming 57, 77–8, 80
carbon monoxide 45, 65
carbon tetrachloride (CCl_4) 19–20, 33, 34, 64, 199
carcinogenesis, sunlight-induced, see photocarcinogenesis and ultraviolet radiation, carcinogenic effects
cataracts 139, 220–2, 229
CFCs, see chlorofluorocarbons
chemical industry, see industry
Chemical Manufacturers Association (CMA) 53, 54, 57
childhood, sunburn in 183
China xiii, 6, 131
chlorine, stratospheric xii, 16, 19–20, 24
  chemistry of ozone depletion 17–18, 30–2
  observed Antarctic levels 54–5
  predictions of future levels 34, 41, 138–9
chlorine dioxide 55
chlorine monoxide (ClO) radical 17, 18, 31, 32, 54–5
chlorine nitrate ($ClONO_2$) 17–18, 30, 31, 55, 56
chlorofluorocarbons (CFCs) xii–xiv, 56, 72, 198–9, 235–41
  applications xiv, 115–16, 235–6
  behaviour in atmosphere 15–16, 17, 20
  global production xiii, 116, 117, 130
  infrared absorption 79
  modelling future effects on ozone layer 33–41, 46–7
  predictions of future emissions 30, 45–6
  properties 236–7
  recycling and disposal 7, 126, 246–7, 255–6
  reducing future emissions 7–9, 10, 229–30, 270
    in European Community 109–13
    role of consumer pressures 244–6, 253–6
    see also Montreal Protocol on Substances that Deplete the Ozone Layer
  role in global warming xiv, 9, 57, 78–9, 80, 82, 91–2
  substitutes 6–7, 119, 237–40, 241
    criteria for 237
    potential environmental impacts 63, 65–6, 80–1, 82, 247
    see also HCFC 22; HFC 134a
climate
  faint Sun paradox 106
  natural variability 88, 94, 95
climate change
  changes in vertical ozone distribution and 45, 47
  frequency of extremes and 105
  future projections 57–8, 77–80, 86, 88–96, 101
  international control measures 9, 95–6, 137–9
  observed 86–8
  at regional level 96, 101
  research priorities 257–63
  vulnerable ecosystems 99–103
  see also greenhouse effect
climate sensitivity 90–1, 93, 94–5, 96
clouds
  feedback effects of global warming 77
  infrared absorption 73, 75
  polar stratospheric (PSCs) 31, 56, 82
coastal ecosystems, effects of climate change 101, 102–3
Cockayne's syndrome (CS) 154–5, 156
  mutation studies 151, 152, 155
  xeroderma pigmentosum (XP) with 150, 154
construction industry 246
consumers
  environmental concerns 251–3
  pressures exerted by 10–11, 243–6, 253–6
Consumers' Association 243, 251
crop yields, effects of UVB radiation 201, 203
cutaneous infections 207–8

De Sanctis-Cacchione syndrome 149
developing countries 124, 131–3, 143, 247, 256
dimethyl ether (DME) 238
DNA
  repair defects in genodermatoses 149–54, 155, 156
  UV radiation-induced damage xi, 147–8, 207, 208

see also ultraviolet radiation, carcinogenic effects
Dobson spectrophotometry 49, 231
drought, future probability 105
DuPont chemical company 239, 240

ecosystems
 effects of climate change 99–103
 effects of increased UVB radiation 197–204
El Chichon volcanic eruption 48, 50, 51, 61–2
El Niño Southern Oscillation (ENSO) 48, 49, 50, 51
energy use patterns 9, 70, 103, 141
environmental pressure groups 243–9
environmental problems
 consumers' views 251–3
 difficulties in controlling 267–71
European Community xiv, 9, 109–13, 129, 141, 270
 ratification of Montreal Protocol 133, 134, 256
 Vienna Convention 111, 121
European Community/European Free Trade Association (EC/EFTA) Task Force 261
European Environmental Bureau 255
eyes, changes in xeroderma pigmentosum 148–9

faint Sun paradox 106
fast-food companies 246, 255
Federal Aviation Administration (FAA) 48
Federation of European Aerosol Associations 255
fibroblasts, photosensitive genodermatoses 149–54, 155, 156, 194
fish larvae, effects of increased UVB radiation 202, 203
fluorine 63
foam packaging materials 238, 246, 255
food supply, global, effects of increased UVB radiation 201, 203
forests
 effects of increased UVB radiation 201
 improving management 9
 mid-latitude, effects of climate change 100, 101–2

fossil fuels 9, 103, 106
freckles, skin 170
Friends of the Earth 243–9

genodermatoses, photosensitive 147–56
The Green Consumer Guide 11
greenhouse effect (global warming) 71–82, 85–96, 105–6
 basic mechanisms 72–6
 control measures 9, 69–70, 95–6, 103, 112–13, 135–9, 141, 269
 equilibrium versus transient responses 77, 90
 faint Sun paradox 106
 human health effects 222–4
 observed temperature changes 86–8, 89
 ozone depletion and 45, 71–2, 81–2
 predicted temperature changes 57–8, 77–80, 86, 88–96, 101
 role of CFC substitutes 80–1, 82, 247
 vulnerable ecosystems 99–103
 see also climate change
greenhouse gases xiv, 45, 103, 118–19
 cooling effects on stratosphere 76, 81–2
 estimates of future concentrations 78, 91–2
 infrared absorption 75–6, 79
 projections of global warming effects 57–8, 77–80, 88–92
 see also carbon dioxide; chlorofluorocarbons; HCFC 22; methane; nitrous oxide

halons xiii, 30, 33, 45, 46, 198–9, 262
 behaviour in the atmosphere xii, 17, 20
 Montreal Protocol provisions 128
 ozone depletion potential xiii, 20
 prediction of effects on ozone layer 47
 uses xiv, 115–16
HCFC 22 xiii, 20, 33, 238, 247
 greenhouse potential xiv, 78, 81, 105
 predicted effects on ozone layer 24, 34
HCFCs 238–9, 262
heterogeneous reactions 30–2, 55–6, 60–1, 258, 259–60
HFC 134a 80–1, 239–40, 241, 247
Holdridge life zones, projected changes 101

hydrocarbon (butane/propane) aerosol
  propellants 236, 237, 238
hydrogen chloride (HCl) 17–18, 55, 56
hydroxyl (OH) radicals 45, 66

ice cover, decreased 77
ice crystals, reactions on 31, 56, 60, 61
ICI (Imperial Chemical Industries) 7, 238–41, 247, 248–9
immune function in xeroderma pigmentosum (XP) 149, 156, 193
immune suppression, UV radiation-induced 156, 167–8, 207–8, 212–13
immunosuppressed patients, skin cancer incidence 212
India xiii, 6, 133, 229
industry 248–9, 269, 270
  consumer pressures on 11, 244–6, 253–6
  development of CFC substitutes 6–7, 111–12, 237–41
infections, cutaneous 207–8
infrared radiation, thermal, see thermal infrared radiation
insulation materials 246
Inter-Governmental Panel on Climate Change 10, 257–8
International Ozone Trend Panel, see Ozone Trend Panel
isotretinoin 156

Langerhans cells 193, 212
latitude
  cataract development and 220–1
  melanoma risk and 171–2, 177–8, 209–10, 213
  ozone column density and 20–2, 50–1
  projected ozone depletion and 35–7, 38, 39, 48
  skin cancer risk and 209–10, 215
London Conference on ozone layer (1989) 10

marine ecosystems, effects of increased UVB radiation 202–3
Mcdonald, J., amplification factor 162
melanoma, cutaneous malignant (CMM) 169–85, 191, 213
  acral lentiginous (ALM) 213
  action spectrum for induction 192
  at different latitudes 171–2, 177–8, 209–10, 213

distribution in populations 170–4
in immunosuppressed patients 212
incidence data 169–70, 192, 214, 215
intermittent exposure hypothesis 174–6
lentigo maligna (LMM) 213
nodular (NM) 213
predicting future trends 176–84, 218–20
  behavioural change and 180, 182–3, 184, 220
  latency of UV effect and 183–4
  mathematical models 188–9
  past time trends and 180–2
  using latitude and calculated UV levels 177–8
  using measured UV levels 178–80
site distribution 173–4
skin type/pigmentation and 170, 213
socioeconomic distribution 173
superficial spreading (SSM) 213
time trends 172–3, 180–2
in xeroderma pigmentosum 148, 213
methane 45
future trends 30, 33
predicted ozone levels and 35, 37, 39, 47, 82
role in global warming 57, 78, 79
methyl chloride (CH_3Cl) 19, 33
methylchloroform 33, 65, 139, 199, 238, 248
predicted effects on chlorine levels xiii, 34
Montreal Protocol on Substances that Deplete the Ozone Layer xii–xiv, 5–6, 115–39, 141–3, 243–4, 256
  articles and programme for implementation 125–35
  assessment and review 10, 133, 137–9, 143, 203–4
  development 123–6
  European perspective 111–12
  future chlorine levels and 34, 41, 138–9
  greenhouse contribution of CFCs and 91–2, 95–6
  predicted effects on ozone depletion 23–4, 34–41, 47, 125–6, 138–9, 142
  rationale 115–19
  shortcomings xiii–xiv, 248
  timetable 137

mutation studies
  in Cockayne's syndrome (CS) 151, 152, 155
  in trichiothiodystrophy (TTD) 151, 152, 155
  in xeroderma pigmentosum (XP) 150–2, 153–4, 156

naevi, benign acquired 170, 183
National Aeronautics and Space Administration (NASA) 8, 48, 53, 54, 57, 231
National Oceanic and Atmospheric Administration (NOAA) 48, 53, 54, 57
National Science Foundation (NSF) 53, 54
natural killer (NK) cells 149, 156, 193
Network for the Detection of Stratospheric Change 57
nitric acid (HNO_3) 17, 56
nitric oxide (NO) 55, 56
nitrogen compounds, odd (NO_y) 17–18, 55–6
nitrogen dioxide (NO_2) xi, 61
nitrogen oxides (NO_x) xii, 31, 64, 198
nitrous oxide (N_2O) 17, 18, 30, 56
  future trends 33, 45–6
  predicted ozone depletion and 46–7
  role in global warming 57, 78, 79
Northern Hemisphere
  observed ozone depletion 21–2, 23, 31–2, 50–1, 138
  observed temperature changes 88, 89
  projected ozone perturbations 35–6, 199
  see also Arctic stratosphere
Noxon cliff 61
nuclear bomb tests, atmospheric 49, 63–4
nuclear energy 9, 141

oceanic thermal inertia 77, 90, 96
ocular cancer 149
optical amplification factor (OAF) 166–7, 216
organ transplant recipients, skin cancer incidence 212
ozone xi
  infrared absorption 75
  monitoring 49, 52, 64, 230–1, 259
  stratospheric xi, xii, 22, 32, 47
  see also ozone layer depletion

total column xi
  observed changes 20–2, 23, 48–52, 230
  predicted changes 35–7
trophospheric xi–xii, 47, 50, 59, 78
vertical distribution, predicted changes 22, 45, 47, 117–18
ozone depletion potential (ODP) xii, xiii, 20, 65, 247, 262
ozone layer depletion 15–25, 43–58, 59–65, 116–18, 198–200
  Antarctic, see Antarctic ozone hole
  assessment of effects 119
  chemistry 16–20, 30–2, 60–1
  heterogeneous reactions 30–2, 55–6, 60–1, 258, 259–60
  consumers' views 252–3
  fact sheet xi–xiv
  greenhouse effect and 45, 71–2, 81–2
  harmful effects 45, 117–18, 200
  human health 139, 147, 161–8, 176–84, 207–24, 229
  terrestrial plants and marine organisms 197–204
  see also climate change; greenhouse effect
  international controls 115–39, 141–3, 203–4, 269–70
  environmental/consumer group pressures 10–11, 243–9, 251–6
  European perspective 109–13, 141
  UK Government policies 5–11
  see also Montreal Protocol on Substances that Deplete the Ozone Layer; Vienna Convention for the Protection of the Ozone Layer
  modelling 27–41, 46–8, 199–200, 260
  accuracy 51–2, 56–7, 123–4
  Cambridge University two-dimensional model 32–41
  one-dimensional models 28, 29, 46, 47
  three-dimensional models 29, 46
  two-dimensional models 28, 29–30, 46, 47–8
  uncertainties 29–32, 41, 59–60, 258
  observed 20–2, 23, 48–56, 230
  predicted 23–4, 32–41, 46–8, 125–6, 138–9
  research priorities 257–63

Ozone Trend Panel vii, 21–2, 23, 31–2, 48–9, 198

packaging materials, foam 238, 246, 255
pentane 238
photocarcinogenesis
  experimental 161–8, 170–1
  in xeroderma pigmentosum (XP) 148–54, 156, 194
photoimmunology 167–8
photosensitive human syndromes 147–56
phytoplankton, effects of increased UVB radiation 202
plants, terrestrial, effects of UVB radiation 200–2
polar stratospheric clouds (PSCs) 31, 56, 82
polar vortex, Antarctic 54, 63

quasi-biennial oscillation (QBO) of stratospheric winds 49, 50, 51, 52

racial differences
  in skin cancer 214–15
  in xeroderma pigmentosum (XP) 194
radiation (optical) amplification factor (OAF) 166–7, 216
refrigeration systems 237, 240, 246, 255–6
research priorities 142, 257–63
  areas of uncertainty 258
  cooperative programmes 260–1
  recommendations for future 258–60
  related areas 262–3
  scientific assessment 262
Robertson-Berger (RB) meters 64, 179, 181, 216, 230–1
rodent ulcer, *see* basal cell carcinoma

satellite measurements of ozone levels 49, 52, 230–1
sea levels
  effects of rising 102
  projected rise 96
sea surface temperatures, global mean changes 87
seasonal variations
  Antarctic ozone hole 52, 53
  future ozone perturbations 35–6, 48
  ozone column density 20–2, 23, 49, 50

skin cancer
  epidemiology 214–15
  in immunosuppressed patients 212
  impact of ozone depletion 135, 139, 216–20
  non-melanoma (NMSC) 169, 210–13
    biological amplification factor 164–5, 167, 216–18
    epidemiology 214–15
    experimental studies 161–8
    optical amplification factor 166–7, 216
    photoimmunological effects and 167–8
    predictions of future trends 162, 164–5, 167, 216–18, 219
  role of sunlight in induction 208–13
  in xeroderma pigmentosum (XP) 148, 156, 211, 213
  see also basal cell carcinoma; melanoma, cutaneous malignant; squamous cell carcinoma
skin pigmentation
  skin cancer risk and 170, 209
  xeroderma pigmentosum (XP) and 194
skin types
  classification 209
  melanoma risk and 170, 213
snow cover, decreased 77
socioeconomic distribution of melanoma 173
soil moisture, projected changes 101, 102
solar keratoses 212
solar radiation
  absorption by Earth and its atmosphere 72–3, 74
  exposure, *see* sun exposure
  see also ultraviolet radiation
Southern Hemisphere
  ozone depletion 52, 53, 199
  temperature changes 88, 89
  see also Antarctic ozone hole
squamous cell carcinoma (SCC) 211–13, 214, 215
  at different latitudes 209–10
  experimental studies 161–7, 171
  predictions of future trends 164, 167, 216–18
  in xeroderma pigmentosum 148, 211

stratosphere xi
Stratospheric Ozone Review Group
  (SORG) 8, 257, 258–9, 260–1
sulphate aerosols 31, 60
sun exposure
  in genodermatoses 147–9, 154, 155
  melanoma development and 174–6,
    183–4, 213
  non-melanoma skin cancer and
    161–2, 210–13
sunburn, melanoma development and
  175–6, 183
sunspot activity 88
  melanoma incidence and 184, 219, 220
  ozone levels and 22, 49, 50, 51, 193
  UV levels and 184, 192–3, 219
suntanning lamps 191
supermarket chains 255
supersonic aircraft, ozone
  depleting effects
  31, 63, 198

T lymphocytes
  mutation studies 151–2, 155
  sensitivity to UV light 152, 153, 156
temperatures
  frequency of extreme 105
  global mean
    future projections 101
    observed changes 86–8, 89
    predicted changes 57–8, 77–80, 86,
      88–96, 101
  stratospheric
    formation of PSCs and 63, 82
    predicted changes 37–9, 40, 60, 76,
      81–2
  see also climate change; greenhouse
    effect
terrestrial ecosystems, effects of UVB
  radiation 200–2
thermal inertia, oceanic 77, 90, 96
thermal infrared radiation
  absorption by greenhouse gases 75–6,
    79
  emission by Earth 72, 73, 74–5
6-thioguanine resistance 150–1, 155
Third World countries, *see* developing
  countries
Toronto Conference on the Changing
  Atmosphere (1988) 9, 137, 257

toxic waste disposal 112
trichiothiodystrophy (TTD) 150, 151,
  152, 155, 156
trichloroethane 78
trophosphere xi, 74
  greenhouse effect 74–6
  ozone in, *see* ozone, trophospheric
tropopause 74

UK Government 244, 248, 256
  policies on CFC emissions 6, 7–9,
    10–11
  policies on global warming 9
  voluntary approach to change 7, 245,
    246, 254
ultraviolet (UV) radiation xi
  carcinogenic effects 161–8, 191–2,
    208–13
  action spectrum 163, 165–7, 192,
    216, 217
  cutaneous melanoma 170–84, 213
  dose-effect relationship 163–5
  in xeroderma pigmentosum (XP)
    148–54, 156, 193–4
  see also skin cancer
  cataract induction 220–2
  effects on living organisms 197–204
    marine animals and plants 202–3
    terrestrial plants 200–2
  ground-level measurements 64, 216,
    230–1, 259
  observed trends 59, 180–1, 191, 230
  sunspot activity and 184, 192–3, 219
  immunological effects 156, 167–8,
    207–8, 212–13
  measurement of active dose 193
  protective effects of ozone xii, 45, 117,
    207–8
  sensitivity in genodermatoses
    147–56
  see also sun exposure
United Nations Environment Programme
  (UNEP) 48, 119
  resource problems 142
  role in Montreal Protocol xiii–xiv,
    10, 123, 125, 133, 243–4
  Vienna Convention xii, 121–2
USA, policies on CFC emissions xiv,
  7, 246, 270

Vienna Convention for the Protection
of the Ozone Layer xii, xiii, 111,
121–2, 142
volcanic eruptions 49, 50, 51, 61–2, 88

warming, global, *see* greenhouse effect
waste disposal, toxic 112
water vapour
  feedback effects on global warming 77, 91
  infrared absorption 73, 75
  stratospheric 54, 78, 82
wetlands, effects of climate change 101, 102–3

Which? 253, 254, 255
World Meteorological Organization (WMO) 48, 119

xeroderma pigmentosum (XP) 147, 148–54, 155–6
  cellular studies 149–54, 193–4
  clinical features 148–9
  Cockayne's syndrome (CS) with 150, 154
  skin cancer in 148, 156, 211, 213

zooplankton, effects of increased UVB radiation 202–3